Art, Cybernetics and Pedagogy in Post-War Britain

This is the first full-length study about the British artist Roy Ascott, one of the first cybernetic artists, with a career spanning seven decades to date. The book focuses on his early career, exploring the evolution of his early interests in communication in the context of the rich overlaps between art, science and engineering in Britain during the 1950s and 1960s. The first part of the book looks at Ascott's training and early work. The second part looks solely at Groundcourse, Ascott's extraordinary pedagogical model for visual arts and cybernetics which used an integrative and systems-based model, drawing in behaviourism, analogue machines, performance and games. Using hitherto unpublished photographs and documents, this book will establish a more prominent place for cybernetics in post-war British art.

Kate Sloan teaches for Newcastle University and the University of Edinburgh, where she was previously Henry Moore Foundation Post-Doctoral Fellow 2015–17.

Routledge Advances in Art and Visual Studies

This series is our home for innovative research in the fields of art and visual studies. It includes monographs and targeted edited collections that provide new insights into visual culture and art practice, theory, and research.

For a full list of titles in this series, please visit https://www.routledge.com/Routledge-Advances-in-Art-and-Visual-Studies/book-series/RAVS

Photography and the Contemporary Cultural Condition
Commemorating the Present
Peter D. Osborne

Digital Art, Aesthetic Creation
The Birth of a Medium
Paul Crowther

Geneses of Postmodern Art
Technology as Iconology
Paul Crowther

Film and Modern American Art
The Dialogue between Cinema and painting
Katherine Manthorne

Play and the Artist's Creative Process
The Work of Philip Guston and Eduardo Paolozzi
Elly Thomas

Bridging Communities through Socially Engaged Art
Edited by Alice Wexler and Vida Sabbaghi

Abstract Painting and the Minimalist Critiques
Robert Mangold, David Novros, and Jo Baer in the 1960s
Matthew L. Levy

Art, Cybernetics and Pedagogy in Post-War Britain

Roy Ascott's Groundcourse

Kate Sloan

Routledge
Taylor & Francis Group
LONDON AND NEW YORK

First published 2019 by Routledge

2 Park Square, Milton Park, Abingdon, Oxon OX14 4RN
605 Third Avenue, New York, NY 10017

Routledge is an imprint of the Taylor & Francis Group, an informa business

First issued in paperback 2022

Publisher's Note

The publisher has gone to great lengths to ensure the quality of this reprint but points out that some imperfections in the original copies may be apparent.

Library of Congress Cataloging-in-Publication Data
A catalog record for this title has been requested

ISBN: 978-1-138-60557-2 (hbk)
ISBN: 978-1-03-233860-6 (pbk)
DOI: 10.4324/9780429468018

Typeset in Sabon
by Deanta Global Publishing Services, Chennai, India

For Product Safety Concerns and Information please contact our EU representative: GPSR@taylorandfrancis.com
Taylor & Francis Verlag GmbH, Kaufingerstraße 24, 80331 München, Germany.

To Roy Ascott

Contents

Figures

Plates

Acknowledgements

Like every book, this one is only complete because of the kindness and support of a whole network of people. It began in the History of Art Department at the University of Edinburgh, which supported me through a PhD and then hosted me for the post-doctoral fellowship that allowed me to write this book. The extraordinary staff there have been supportive and inspiring, particularly Richard Williams who has over a period of years provided enthusiasm, advice and keen insight. I would also like to thank Andrew Patrizio for his kindness and excellent and very practical supervision and Carol Richardson for her support during my post-doctoral fellowship, as well as Jill Burke and Neil Mullholland.

I am also enormously grateful to the Henry Moore Foundation who granted me a two-year post-doctoral research fellowship which allowed me that increasingly rare commodity of time to transform part of my PhD into this book. Above and beyond the funding, this organisation supports researchers to build connections and to actively share their research. I benefitted immeasurably from my time working with HMF, including having the opportunity to put together an exhibition of Ascott's work. Special thanks go to Jon Wood, Lisa Le Feuvre and Kirstie Gregory, who each in their own way inspired me and made me ambitious for my project.

I am so grateful to my loving and supportive family, all of whom have urged me on through this project and didn't let mere details like new motherhood and back surgery stand in the way of this book. My parents and my brothers have been unwavering in their support, which has sustained me through several years of work. I also want to send love and thanks to Tony O'Malley, whose genuine pride and interest in my work – as well as his outrageous teasing – have been a source of great motivation.

Finally, I need to thank Roy Ascott, without whom this book would have been impossible. He is a continuous source of surprise and inspiration and as I wrote, I also learned in ways I didn't anticipate. This book is dedicated with gratitude to Ascott, an extraordinary artist and educator. I hope that others might find inspiration in these pages too.

Dialogue (Introduction)

In the immediate post-war years we entered into what the British artist Roy Ascott describes as 'an age of dialogue'.[1] As the country emerged from the shadow of World War II, the many innovations in science, engineering and technology that the war had produced began to disperse into industry. It was quickly evident that a new age for technology had started. Biology was engineered and machines were replicating living systems in increasingly sophisticated ways, prompting the development of the field of cybernetics. Ascott was one of the first cybernetic artists in the UK and from this early engagement, he developed a fascination with the communication potentials of technology that would span the seventy years that followed.[2] At the point of writing, Ascott is in his seventh decade of practice and remains one of the most influential and innovative artists working with technology today.[3] This book argues that in his early career, Ascott created and enacted a 'cybernetic manifesto' that constituted an innovative and influential contribution to the field of art and technology. The book will draw out the development of a sequence of ideas that has been the basis of his lifelong working practice. It argues that Ascott made a sustained, and often overlooked, contribution to British art in the period and that his cybernetic pedagogy influenced a generation of artists that would go on to experiment in the then undefined field of multimedia. Ascott's work in the 1950s and 1960s brought together biology, abstraction, cybernetics, interactivity and communication systems. It stands as the most perfect exemplar of his own term 'dialogue', in that it embodied the power and potential of interdisciplinary exchange.

When Charlie Gere wrote in 2002 that 'Cybernetic and computer art was [after the sixties], rightly or wrongly, regarded as marginal in relation to both the traditional art establishment or to avant-garde art practice', he identified a problem that has yet to be fully resolved.[4] Ascott's career has certainly not had the scholarly or critical attention it deserves and this is in a large part due to the marginalisation, ring-fencing and persistent misunderstanding of the place of cybernetics in twentieth-century art history.[5] We can partly attribute the deficit of scholarship on cybernetic art to the difficulties of displaying, recreating or maintaining cybernetic art and exhibitions, rooted as they are in defunct or erratically functioning technologies. However, the problem runs deeper than this, in that the complex area of art and cybernetics is still most often understood an adjunct to conceptual art. In truth, many of the questions of systems, networks and environments of communication which formed the core of cybernetic art were also vital to developments in other areas of conceptual art. I argue that cybernetic art was simply the most direct confrontation with the interrelated technological and social change that had provoked a seismic shift in the visual arts: from object to concept.

Cybernetics was a milestone in the history of technology and the questions that arose from these early theories and experiments in communication systems still have relevance today. Ours is a cybernetic world, in which communication between machines and mankind has spread into every facet of contemporary life. In light of this, this book forms part of a revival of interest in documenting and scrutinising cybernetic art as a 'history of the present' that has been gathering pace over the last half decade across the world.[6] More specifically, while giving a full account of the overlapping interests that contributed to Ascott's development, the book will reconstruct the distinctive and often eccentric character of British cybernetic art. For Ascott, cybernetics was a lived experience and through it, the dawning of our network age was visible and tangible. As one of a generation of young artists that made their transition from the military to the art school studio, he witnessed first-hand the sophisticated communication technologies that had changed warfare forever. After a war that had prompted collaboration between engineers, scientists, psychiatrists and mathematicians, the communication technologies that had been invented during the war dispersed and adapted into other social and economic applications. At the same time in Britain, the recession and the physical damage wrought by the war created a state of confliction that hovered between trauma and optimism. Given that this collective trauma was tied inextricably to technologies such as atomic weapons, it is little surprise that the technological optimism of the period 1945–70 is often cast in shadow. However, this optimism was a defining feature of this period of rich invention and it was not only present in the sciences, but also in the arts. Indeed, collaborations, residencies and other forms of partnership between the sciences and the arts were frequent and in the USA often well-funded.[7]

In Britain, economic depression limited the scope for well-funded collaboration, but it did not limit the thirst for collaboration and experimentation. Indeed, artists and scientists alike worked with what was available. While British cyberneticists such as Gordon Pask and William Ross Ashby repurposed war detritus such as bombsites in their experiments, the surge in collage, welded metal and constructivist sculpture in post-war Britain also saw materials repurposed. One of the most interesting facets of this period of British history was the way in which a general thirst for the new, for an optimistic future and for technological change grew in an age of rationing and scrap and against a backdrop of ruined cityscapes marred with the yawning absences of missing buildings. It is easy then, to read the art of the post-war era in the context of social and cultural trauma. However, Ascott and other artists that worked with technology in 1960s Britain saw it as an age of optimism, discovery and new possibility for human communication. Ascott saw this new age of communication in close embrace with the visual arts and he predicted that it would transform the conceptual framework of a discipline so long defined by objects.

We can partly attribute this shift towards a more communication-focused mode of working to the fact that artists and designers explored the possibilities of new technologies through play – that is, they often used technologies in ways outside of their intended function, leading to fruitful experiment that in turn opened out new possibilities. Play was also one of the cornerstones of art and design education in the immediate post-war decades of 1945–70. This phenomenon related to the developing field of literature on the psychology of art, particularly in relation to the intuitive approaches of children.[8] The theorist Herbert Read was broadly influential upon British art education, particularly with regards to the experimental teaching that took place at Leeds College of Art and King's College, the University of Durham, where Ascott himself studied and subsequently worked. Read argued for a broadly individualised approach to art teaching, in recognition of the fact that art was 'The earliest and most exact index to the child's individual psychology'.[9] Read's approach certainly informed teaching approaches in the Basic Design movement, in which tutors must not be '[...] a dictator, but rather a pupil more advanced in technique than the others, more conscious of the aim to be achieved and the means that must be adopted [...]'.[10] Game play was often an important constituent of the experimental projects and exercises that Victor Pasmore and Richard Hamilton taught at King's College during Ascott's time there, particularly in exercises that employed throwing or other randomising elements in the creation of abstract imagery.[11] These playful approaches in art education placed an increasingly loaded emphasis on process over product, which is in itself another indicator of Ascott's 'age of dialogue'. Certainly, Ascott's formative experiences at King's College were a profound influence upon his own his practice and pedagogy in the 1960s.

As well as his art school training, Ascott's other formative experience was the two years of military service he had completed before commencing his degree studies in 1955. As a fighter control officer, Ascott had worked in the singular and secretive environment of the military bunker, tracking flights and assessing threats across the expanse of the North Sea. The bunker was essentially a cybernetic environment, a communication system between man and machine. The majority of artists of Ascott's generation had either served in the war or completed military service after the war and this may be one of the most important and least recognised influences upon the period of fertile development in the arts that followed the war. Ascott was one of the first artists to grasp the significant impact that communication technologies were to have for the visual arts. He recognised that cybernetics, information theory, computing and network theories represented a change to human communication that was not about hardware. Rather, it was about knowledge and meaning as a continuous and unstable process of exchange, and art production as a system that extended out into time. In this regard, Ascott's conceptualisation of cybernetics as a model for art practice and teaching was without parallel.

When Ascott discovered his first cybernetic books while browsing the stacks at King's College Library then, he had already experienced the curious new world of communication between machines, and between machines and humankind, directly. The Fighter Control bunker was a cybernetic environment in which information from radar screens was replicated in analogue form using coloured wood and plastic markers pushed into place on the control room map. Ascott and other officers watched remotely, from tiered seating that gave them a bird's eye view of the map below. In the artificially lit darkness, these activities had the concentrated performativity of a theatre performance. Indeed, in his early years as an art student he produced a design for a fantastical electric theatre set-up he called *Mobile Colour Symphony Hall* (Figure 0.1).

For a young man who had already developed interests in both theatre and the visual arts, it was an arresting experience. Stepping back in time from the desensitised worldview of the thoroughly technologised twenty-first century, we can appreciate the strangeness and the newness of communication technologies during and after World War II: the static, the bleeps and the wavering pulses, the lines and points of light upon screens. What Ascott discovered in cybernetics was the theory of what he had witnessed: a new realm of communication in which information was disembodied into *signal* and *sign*. In order to understand the relevance of this to Ascott's art practice and pedagogy, we must turn to some of the most important research taking place in British cybernetics in the period.

One of the important influences for Ascott was the British cyberneticist William Ross Ashby, who had, in 1948, invented a 'homeostat'– a system that could respond to changes in conditions and maintain its own stability. Ashby's 'Homeostatic System' is considered the earliest device to have been capable of adapting to its own environment in this way (Figure 0.2). In 1951, Ashby wrote of the important change that had taken place in social attitudes to what a machine was and what it might become:

> It has become apparent that when we used to doubt whether the brain could be a machine, our doubts were due chiefly to the fact that by 'machine' we understood some mechanism of very simple type. Familiar with the bicycle and the typewriter, we were in great danger of taking them as the type of all machines. The last decade, however, has corrected this error. It has taught us how restricted our outlook used to be; for it developed mechanisms that far transcended the utmost that had been thought possible, and taught us that 'mechanism' was still far from exhausted in its possibilities. Today we know only that the possibilities extend beyond our farthest vision.[12]

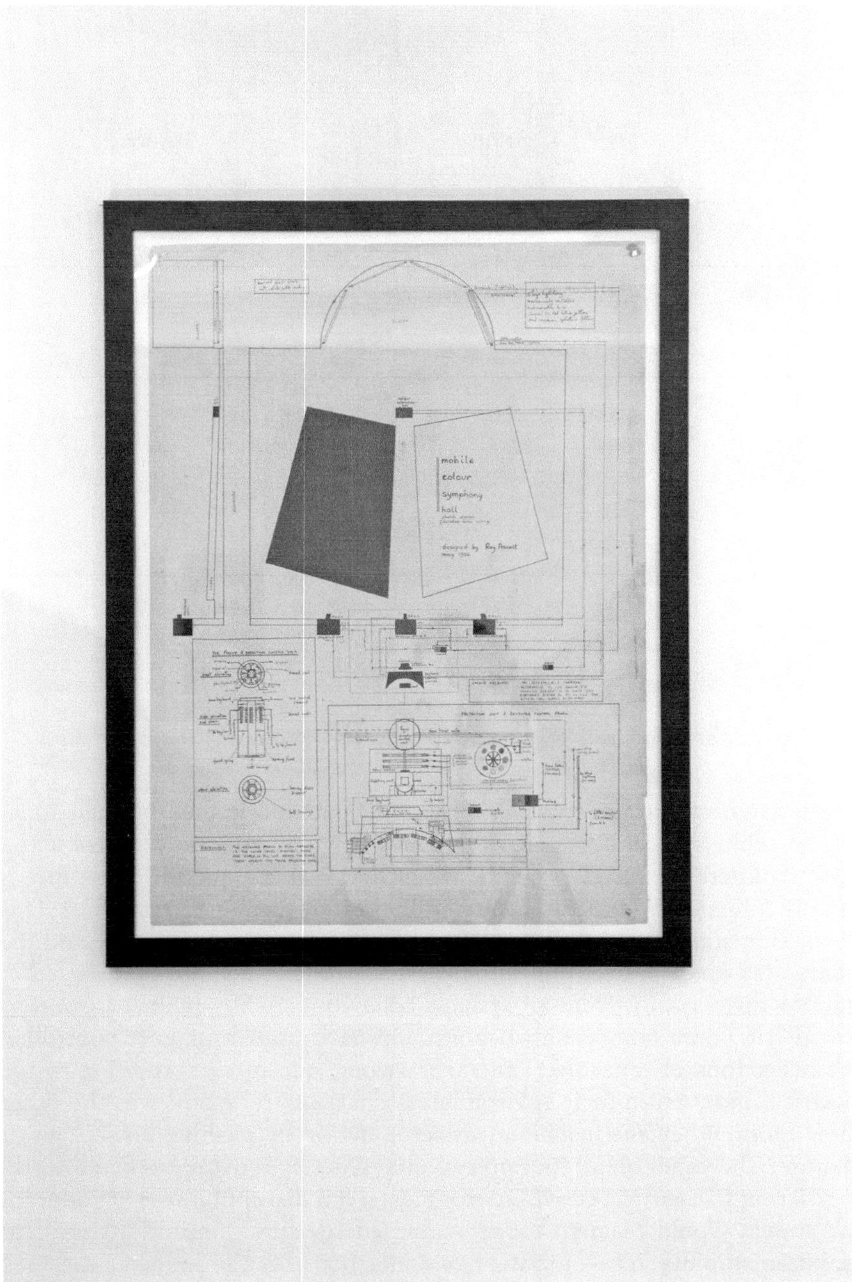

Figure 0.1 Roy Ascott (1956) *Mobile Colour Symphony Hall.* Ink on Paper. Photograph © Roy Ascott

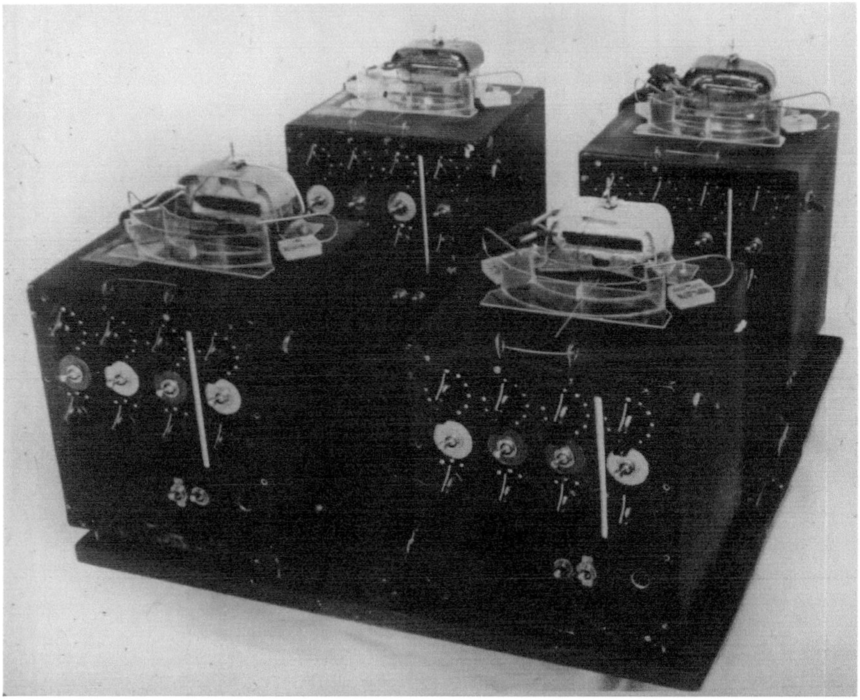

Figure 0.2 Ashby's homeostat. Photograph courtesy of the W. Ross Ashby Estate

The typewriter and the bicycle are both defined and limited by their function, machines as tools. As Frederic Jameson would later write, there was a distinct difference between the machines that had enamoured the Futurists into '[...] relatively mimetic idolatry' and the new machines of media.[13] This distinction started with the systems technologies of World War II, and the homeostat was a perfect philosophical machine that held no purpose other than to further knowledge of systems behaviour. Ashby built it from four Royal Air Force bomb control units. On each unit, a magnet controlled the deflections of a needle. The vanes swung side to side, reacting to the electrical input from their environment. That 'environment' was the other three units. When the machine was set in action by pushing a vane out of position, the vanes on all four units would react by moving back and forth, in reaction to their respective environments. If a disruption was introduced, the system would react to the changing energy flows and adjust itself to maintain stability. The output of one unit becomes the input of the next unit. Ashby described this self-regulation as 'ultrastability'.[14]

Homeostasis is the ability of a system to regulate itself in order to remain stable. Norbert Wiener later described this in relation to the human body, writing cheerfully that: 'We can continue to live in the very special environment which we carry forward with us only until we begin to decay more quickly than we can reconstitute ourselves. Then we die.'[15] Reconstitution is essentially homeostatic in that we need to maintain the core stability of the body's functions – its temperature, the beat of the heart, the health of the digestive tract, the flow of blood – in order to survive. The system breaks down if there is a disruption that causes more damage than our bodies can repair and regulate. Ashby's homeostat revealed that a system will work to maintain stability, and hence to continue fulfilling its own function. Ashby observed: 'The "homeostat" [...] has demonstrated that the deliberate introduction of randomness into a machine, with suitable corrective feedback, can give a flexibility of action and power of adaptation not seen before.'[16]

So, the homeostat was a system, decentred and defined by the flow of communication between the four units which had been repurposed from war detritus. Its main characteristics were responsiveness and adaptability, which in turn led to a new complexity:

> The classical method for studying a machine is to identify the actions of the parts and then construct, step by step, a theory of how they will act in unison. But when the number of parts is increased without limit this method becomes totally impossible.[17]

Considering this, it is easy to see why Ashby began to believe in the theory of the artificial brain; if a rudimentary system such as this could respond and adapt to survive, then the science fiction dream of machines that had the dynamism of the human mind might yet become reality. Ashby observed that '[...] man does not think logically – he thinks dynamically'.[18] Dynamism, then, was the most important requirement in the production of a 'thinking' machine. By 1960, Ashby was predicting a future in which industry would be automated:

> In the spate of plans and directives issuing from it we might hardly notice that the automatic valve-making factories are to be moved so as to deliver directly into its own automatic valve-replacing gear; we might hardly notice that its new power supplies are to come directly from its own automatic atomic piles; we might not realise that it had already decided that its human attendants were no longer necessary. How will it end? I suggest that the simplest way to find out is to make the thing and see.[19]

Cyberneticists certainly predicted the dilemma of automated industry that is the primary economic change of our own age.

The homeostat's relevance to Ascott's art practice is quite direct in that it was decentred, a system rather than an object. Given that in the 1960s and the 1970s systems approaches dominated in the rise of conceptual art, the notion of the cybernetic system as information that adapted in transmission was a powerful one. It should also be noted that the first formal theories of information were created in the realm of cybernetics, engineering and mathematics, often using compelling diagrammatic descriptions of information flow, such as C.E. Shannon's 'A Mathematical Theory of Communication'.[20] For Ascott these influences were direct in that he engaged directly with cybernetic research, attended events and conferences, collected literature and built relationships with scientists including Gordon Pask. However, the transmission of language from communication theories in the sciences to the arts is also evidence of a broader shift in attitude to communication in the arts, resting upon the principles of transmission, signal and adaptation.

In order to understand the intuitive and experimental magic of early cybernetics that proved so transformative for Ascott, we need only look to two of the most creative scientists of the twenty-first century: Gordon Pask and Stafford Beer. In 1956 or 1957, they grew an ear. To be more accurate, they built a machine that exhibited the intent to grow an ear in order to respond to sound. The two men had been preoccupied with similar ideas during the first part of the decade, although Beer had diverged into '[...] attempting to use populations of biological organisms (such as the water flea Daphnia) to compute complex functions', while Pask had been attempting the same using electrochemical set-ups.[21]

The pair shared a restless intellectual curiosity which drove them into an overnight experiment that was designed to probe Ashby's theory of ultrastability, particularly how change and disruption might influence a stable system. As Peter Cariani has described, Pask had been preoccupied for some time with creating '[...] self-organizing devices' that '[...] passed current through metallic structures (iron, tin, silver) immersed in an acidic milieu (sulphuric, nitric acid), often in capillary tubes or dishes'.[22] Cariani continues:

> The potential complexity of the behavior of these electrochemical assemblages was well appreciated by those familiar with the 'iron-wire' neural models that had been around since the turn of the century. These physical models were capable of astonishingly nerve-like properties.'[23]

Their machine consisted of electrochemical dishes. Pask had placed obstructions into the dishes, and the system had grown threads over them in order to overcome the disruption to their system. If the threads were cut and the current reintroduced, the threads regrew, bridging any gap between the anode and the cathode. In 'Gordon Pask: His Maverick Machines', Jon Bird and Ezequiel Di Paolo describe this experiment, which Stafford Beer remembered as 'the most important and indeed exciting of my personal recollections of working with Gordon'.[24] According to Bird and Di Paolo, Beer recalled that Pask had suggested:

> 'Suppose that it were a survival requirement that this thing should learn to respond to sound? If there were no way in which this 'meant' anything, it would be equivalent to your being shot. But this cell is liquid, and in principle sound waves could affect it. It's like your being able to accommodate to a slap, rather than a bullet. We need to see whether the cell can learn to reinforce successful behaviour by responding to the volume of sound.' [...] It sounded like an ideal critical experiment.[25]

Bird and Di Paolo explained that 'The electrochemical system is not just electrically connected to the external world: threads are also sensitive to environmental perturbations such as vibrations, temperature, chemical environment, and magnetic fields', meaning that these 'arbitrary disturbances' could be a 'stimulus for the system, especially if they cause a change in current supply'.[26] It was nearly morning as they set a microphone to pick up the sounds of the traffic on Baker Street, Gordon Pask studying the cell. According to Beer, Pask turned to him and uttered the words: 'It's growing an ear.'[27] So, Ashby's homeostat responded to disturbance in order to maintain stability; Pask and Beer created a machine that adapted in response to newly introduced elements. A mass of fine wire, iron filings and chemicals with a current running through them arranged itself into a funnel in order to 'hear' the vibrations and rumbles of Baker Street traffic. In a later account of this episode, Pask commented: 'The ear, incidentally, looks rather like an ear.'[28]

This extraordinary little experiment is a window into the living, complex and unpredictable character of early cybernetics, to which Pask and Beer were important contributors. Their ability to articulate the important questions produced by new research was coupled with an innate understanding of the capacity machines had to surprise; in this experiment, they strayed from the perimeters of an experiment testing the closed variables of a homeostatic system to ask 'what if [...]?' What if, in this case, we introduce a new variable of sound? The outcome was, the iron filings that would build bridges between cathodes and anodes in order to protect the function of the machine, reconfigured to 'read' sound vibrations.[29] It grew an ear. Pask and Beer were demonstrating systems adaptability, but they also demonstrated the extent to which systems behaviour was beyond comprehension or control. Even rudimentary electrochemical systems could learn, evolve and change; machines exhibit behaviour. This generation worked intuitively and discovered that as they did, so did their machines. Their discoveries left them only with more questions, more possibilities and less understanding, in that they were learning through doing, even through play. The only way to see what a particular system would do was to build it and to test it. Conversely, breaking it down into its constituent elements did not advance any understanding of how it operated. It is also worth noting the increasing interest in how a machine *behaved* rather than simply operated that began with early cybernetics and extended into the next two decades.

Machine 'behaviour' is a compelling term that bestows volatile life upon the machine. Ascott was also to employ the term behaviour in relation to his art as early as 1960, by which point he was preoccupied with the 'behaviour' of form in art, biology and in the machine.

It is evident then, that play was a tactic applied in several disciplines in the immediate post-war years. There were many contributing factors including what Ashby described as the 'black box' problem, in that the only way to learn about a system was to experiment with it, not to take it apart.[30] This mode of working was naturally playful because it was indirect, always pursuing possibility. In this period of cross-fertilisation between disciplines, an open-ended approach to experimental play evolved. I have already noted the influence of child psychology upon visual arts education, but beyond this, play was emerging as a mode of practice in a period of rapid social and technological change. Given the simultaneous period of rapid change in the visual arts and in technology and engineering, the themes of systems, play, experiment and communication methods were not simply chance. In both cases, they were new tactics for a networked world, while dealing with a technological potential that was both unknown and immense.

If we view cybernetic theory from what was Ascott's position – that of an ambitious and talented young artist who was already interested in communication structures and symbols – then it becomes a compelling narrative of responsive change, with a strange and captivating language. For Ascott, it also captured the curious performativity of both art and technology, in which the viewer-participant took a self-conscious part in a theatre of objects; this was true of both the Fighter Control bunker and the exhibition space. Beer recalled that he, Pask and others working in the realm of cybernetics had quickly begun to make what was a reformatory realisation for the sciences:

> [...] whereas it had been made clear by Heisenberg in physics decades earlier that the observer affected the system observed, it was now obvious that the observer was not only a part of a cybernetic system as being 'in the loop', but intrinsic to it.[31]

This shift towards a more reflexive understanding of scientific experiment certainly chimes with the reflexive experiments that took place in the visual arts in the 1960s – the rise of performance, of shock and of planned audience interaction from Experiments in Art and Technology's *9 Evenings* to Yoko Ono's performances with their reliance on audience participation. It is also important to note that Gordon Pask was a good friend of Ascott's, who had given a lecture at Ealing to Groundcourse students and staff, collaborated with Ascott on various projects and on a memorable occasion, translated a book of Russian cybernetic theory for him over a bottle of whiskey.[32] Pask was also an early contributor to the field of cybernetic art, creating works of sculpture including *A Colloquy of Mobiles* and his theatre-light show *Musicolour*.

In light of this, I suggest that reflexivity was both a cybernetic concept and a cultural one, that extended from the machine and beyond it, to express a new sensibility of adaptive communication for a society that had come to accept instability as permanent. Indeed, cybernetics offers us an alternative narrative to the crisis theories that are so embedded into our understanding of post-war culture, because cybernetics was innately positive and future-facing. Playfulness was a tactic born out of the best kind of ignorance, in that the great and undeveloped potential of communication technologies was a profound and magical new territory for science, and also, as Ascott had identified, for art.

Ascott and cybernetic art in Britain

When Ascott began to explore cybernetics, he was attracted to the field not only for creative reasons but also reasons of ambition; he needed a point of difference to identify him as an artist in a mass of competitive talent and he decided he could be the 'cybernetics guy'.[33] In 1961 the concept of cybernetics was still esoteric enough to set Ascott apart. It was a new realm for art. There was strangeness and novelty of the phenomenological experience of machine communication – static, bleeps and signals. For Ascott, the profundity of cybernetics lay in its expression of communication structures and codes, but he also recalls with clarity the weirdness of electronic noise in the early stages of communication technology.[34] The language of communication theory – particularly the words and diagrams of C.E. Shannon – was flexible enough to describe different kinds of communication, including the visual.[35] In 1950 Norbert Wiener, the founding voice of American cybernetics had written:

> [...] Society can only be understood through a study of the messages and the communication facilities which belong to it; and that in the future development of these messages and communication facilities, messages between man and machines, between machines and man, and between machine and machine, are destined to play an ever-increasing part.[36]

Ascott saw the communication theory embedded in cybernetics as a flexible and analogical model for visual communication and at the same time, he recognised that computer technologies would rapidly transform human communication. For Ascott then, cybernetic art was essentially a communication model and it was not defined by the use of computers or other communication technologies.

In the field of art history, cybernetic art is often misunderstood as a brief and early incarnation of computer art. In the UK this problem originates from the late 1960s, when cybernetic art became more prominent and less flexible in its outputs. There is little formal recognition of the active interdisciplinary work on art and cybernetics which took place in Britain from the beginning of the 1960s. This issue has been discussed by Edward Shanken, most fully in his essay 'Cybernetics and Art: Cultural Convergence in the 1960s', in which he explores Ascott's practice in connection with the ideology of cybernetics.[37] However, the problem goes beyond formal recognition of the practices of Ascott and other artists of the period. The influence of communication technologies upon the visual arts in the post-war period did not result in a singular, cybernetic mode of practice. It infiltrated the visual arts in myriad ways and, as I will progressively argue, played a key role in the confluence of conceptual approaches that emerged in the 1960s and 1970s.

The most widely known manifestation of cybernetic art in Britain was Jasia Reichardt's exhibition *Cybernetic Serendipity*, which took place in 1968 at the Institute of Contemporary Art (ICA) in London. A vocabulary of cybernetics took the form of an index in the catalogue for the exhibition *Cybernetic Serendipity*, an exhibition that took place at the ICA from 2 August to 28 October 1968. The exhibition poster had a subtitle reading 'the faculty of making happy chance discoveries by means of control and communication machines, both human and electronic'.

The Studio International special edition for the exhibition included a glossary of terms and concepts for the reader, indicating that many of the terms, including *feedback, interface, software* and *bionics*, were not yet common parlance. The field of cybernetics emerged in the 1940s – it evolved alongside mechanical engineering because it was a necessary layer of engagement with the issues of systems, feedback and control that sophisticated new technologies presented. Its etymological root is from the Greek (*kybernētēs*), meaning 'steersman, governor, pilot, or rudder' – an indication that the core issue in cybernetics is that of controlling or giving direction to complex systems, both human and technological. This definition was included by Jasia Reichardt in her press release for the *Cybernetic Serendipity* exhibition, and then subsequently used by Shanken while forging a better definition of cybernetics in relation to Ascott's work.[38] This etymology gives us an understanding of the first cybernetic principles: it was about the human control of technology, as well as the potential of technologies with the sophisticated abilities of the human brain. Its objectives engaged with the reality of the post-war technological world, with humankind at the helm of more complex technological systems than ever before. In the introduction to his 1960 book *Design for a Brain,* William Ross Ashby wrote that:

> [...] a system can be both mechanistic in nature and yet produce behaviour that is adaptive. I hope to show that the essential difference between the brain and any machine made yet is that the brain makes extensive use of a method hitherto little used in machines. I hope to show that by the use of this method a machine's behaviour may be made as adaptive as we please, and that the method may be capable of explaining even the adaptiveness of Man.[39]

Cybernetics theorised the wonder and terror of science fiction dreams of automata, but Ashby, Weiner, Walter and others presented the idea of a mechanical brain as possibility, not as fantasy. Simultaneously, the fields of behaviourism and organisational science both grew from the same ground as cybernetics and targeted contemporary interests in recognising and channelling patterns of human behaviour, particularly in groups and networks such as factory workers.[40] Ascott came across the writing of B.F. Skinner in the same period as he discovered cybernetics, something that had a particularly prominent influence on his subsequent teaching.[41] The politics of power that informed behaviourism and the application of behaviourist tactics in the visual arts are discussed at several points in this book. For now I will note that even in early cybernetic books, the authors demonstrated an awareness of issues such as control and agency, as well as the potential economic and social ramifications of machines supplanting people.

Ascott's work was deemed unsuitable for *Cybernetic Serendipity*. The next issue of *Studio International* included a letter from Ascott to the editor in which he affirmed his own place as '[...] the artist responsible for first introducing cybernetic theory into art education in this country (Ealing 1961) and for having disseminated the concept of a cybernetic vision in art through various art and scientific journals in recent years'.[42] Despite the fact that Ascott's pedagogy and practice were rooted in cybernetics, the exhibition concentrated on cybernetic art of a more technological – or mechanical – nature. *Cybernetic Serendipity* suffered from what Rainer Usselmann has described as a 'pseudo-progressive message, wrapped up in a fun-fair of blinking, hooting robots'.[43] This highlights an interesting issue – while the look of the art displayed as part of the exhibition was superficially more technical than that created in Groundcourse by a collective of staff and students, it did not connect with the principle of cybernetics in a sophisticated way. While the exhibition stands as a good measure of how prominent cybernetic theory had been in the art scene of the 1960s, by defining the boundaries of cybernetic art in such material and technical terms its broader reach and influence were diminished. Ascott was understandably frustrated by *Cybernetic Serendipity*, primarily because it negated the meaning of cybernetics and also because of the personal slight in having his work ignored. It is also evident from the distance of half a century that *Cybernetic Serendipity* marked the end of the high years of cybernetics, as well as cybernetic art. The field of cybernetics did not collapse though; it dispersed into other applications in step with the rapidly accumulating developments in computing. Similarly, artists working with technology moved on in step with technological change.

Biological convergences: behaviourism, cybernetics and military systems

Another important influence for Ascott at the beginning of the 1960s was behaviourism, particularly the work of B.F. Skinner. Convergences of cybernetics, biology and behaviourism during World War II later led to surprising bio-technical models of control in the visual arts in the post-war years.

Groundcourse was the most fully realised example of this phenomenon, set apart by the fact that Ascott had designed the course with direct reference to cybernetics and behaviourism, including the work of Skinner, who was the best-known behavioural scientist of the era. While Skinner's popular works were to become widely read and to attract interdisciplinary interest, his comparatively overlooked research dating to the war years serves as an apt example of how far cybernetics and behaviourism were intertwined.[44] During World War II, Skinner was passionately immersed in *Project Pigeon*, later labelled *Project Orcon*, a contraction of 'organic control' (Figure 0.3).[45] In an episode of US military research that fused behaviourism, cybernetics and biology, *Project Pigeon* was Skinner's attempt to develop a pigeon-controlled missile, with funding of $25,000 provided by the National Defense Research Committee. In 1938, Skinner had published *The Behaviour of Organisms: An Experimental Analysis*. He had argued toward a science of behaviour, exploring the difficulties that had restrained scientists from working in this area of systems biology. He related a series of controlled experiments which used positive and negative stimuli such as food deprivation or supply, evidencing the fact that behaviour could be effectively modified using systems of control. Then the war had started, and he began to consider the practical application of behaviourism in the military.

Project Pigeon was built upon the premise that the behaviour of an organism could be conditioned and hence controlled; in this case the pigeon was to be conditioned to form part of a weapon. Skinner proposed a new missile control system in which a lens would be secured to the front of a missile and the image projected upon a screen inside. The pigeon would be trained to recognize the target on this screen, and would respond by pecking at it. If the pecks remained at the centre of the screen, then the missile would continue to fly straight; if the pecks landed off-centre then it would

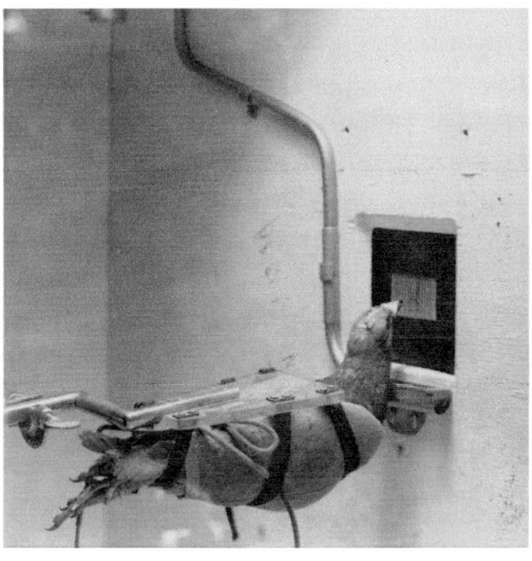

Figure 0.3 The pigeon-guided missile experiment. Photograph courtesy of the B.F. Skinner Foundation

trigger a change in the missile's flight controls and it would change course. The research showed surprising promise but funding was eventually withdrawn. More surprisingly, the pigeons were not the problem; it was the reliability of the servomechanisms upon which this pigeon-missile control system operated. In his article 'Engineering Behaviour: The Conditioning of B.F. Skinner', James H. Capshew described how '[...] the problems were mechanical rather than psychological', in that '[...] the development of mechanical linkages for translating the signal into steering movements proved difficult'.[46] While Skinner put his behavioural science to practical application during the war, it is important to note that his sustained research was directed towards understanding human behaviour; as noted, his white rats were chosen for their similar 'sensory equipment' to that of humans. In his exploration of conditioned responses, he highlighted possibilities for modifying human behaviour by extension. Capshew credits the war years as the period in which Skinner saw that his '[...] system of behavioral science could engender a system of behavioral engineering'.[47] While the humble pigeon employed within Project Orcon was treated as an expendable entity, Skinner's design – the pigeon as part of a mechanical-biological control system – was also an apt illustration of the new warfare of human-technological systems. Capshew noted that the project was instrumental in making Skinner redirect '[...] his thinking toward the possibilities of behavioral engineering'.[48]

The most complex system with which the cyberneticist had to work was the human brain. N. Katherine Hayles recounted early transitions in cybernetic thought in her essay 'Boundary Disputes: Homeostasis, Reflexivity, and the Foundations of Cybernetics'.[49] She outlined the discourse provoked at the Macy conference series in the United States, which had brought together the cybernetic community for discussion and dissemination, between 1946 and 1953. Hayles describes how at the sixth conference, John Stroud used the example of an operator with a radar-tracking device on one side and an antiaircraft gun on the other to identify the complexity of the living element of any system: 'The Gun operator, Stroud observed, is "surrounded on both sides by very precisely known mechanisms and the question comes up, 'What kind of a machine have we put in the middle."'[50]

Hayle recounts how Stroud saw man as an 'input/output device' in the centre of these circuits, a position that was then contested by Frank Fremont-Smith:

> 'Certainly [man] is never only in between two machines when you are studying him because you are the other man who is making an input into the man. You are studying and changing his relation to the machines by virtue of the fact you are studying him.'[51]

This perspective challenged the idea that the man operating the machines could be considered a stable element of the system. Rather, he was a changing entity and the act of studying him changed the system – in other words, any observer was also part of the system.

In the early years of cybernetics, theorists frequently aligned the brain with the machine. Norbert Wiener later observed that in this model of thinking, 'the synapse in the living organism corresponds to the switching device in the machine'.[52] The genesis of cybernetics in Britain was notably distinctive in its focus on modelling human and animal behaviour, much influenced by the fact that the new science was advanced by biologists and psychiatrists. There are then, sharply relevant parallels between Skinner's wartime experiments in 'organic control' and the cybernetic project of understanding and modelling the brain as the most complex mechanism within a system. The living component in each case is the least stable and the most mysterious part of an engineered system: that is, the behaviour of the living had to be modified or controlled in order for the system to successfully operate. In this way, Skinner's experiment in conditioned control and its application in wartime research illustrates the convergence of biology, technology and systems in the sciences during the war; an indication of the ground from which interdisciplinary systems theories evolved. Retrospectively, we might place behaviourism within the broad cybernetic spectrum of scientific, engineering and cultural outputs and in doing so, emphasise the humanity of cybernetics, its critical engagement with living systems. The converging interests of behaviourism and cybernetics share the same point of departure: that of systems of behaviour and how they might be both analysed and controlled.

While post-war developments in behaviourism and social control connect profoundly to early cybernetics, the point of difference for behaviourism is plain. The cybernetic preoccupation with control was largely theoretical, manifested in experiments that harnessed and controlled increasingly complex systems of technology. Behaviourism, on the other hand, was largely a social theory, focussed on harnessing and controlling humanity. The white rat was a symbol for the human subject; any kind of behaviour modification programmes constitute a problematic power relationship in which an organisation – or an individual – might control or alter the behaviour of others. While in tests, Skinner's servomechanisms proved more volatile than the trained pigeons, as we know, the living are unpredictable. A body might spasm, a mind drift. A heart might stop beating. The most established programmes of behaviour modification might be subverted, ignored or undone because the living are creators and destroyers of systems of control. Ascott's pedagogy operated on this foundational principle – the fact that we are possibility and as long as we recognise the patterns that make up our *selves*, we can subvert patterns and systems and, in doing so, exact change.

Despite its abandonment, Skinner's *Project Orcon* was an arresting model for an organic–mechanical control system, developed during the same war that saw the role of military personnel become increasingly systematised. Their actions and reactions were shaped by how they responded to information; and enacted through, or by, machines. One might argue that there is little distance between the pilot and the pigeon here; both are caught in an established system of control, their behaviour and responses shaped by training. In this light, the post-war influence of behaviourism upon British society and culture stands in stark contrast to the organisational analysis of the interwar period. Interwar organisational study was in some regards the precursor of behavioural science. It focused on employee satisfaction and social relations, in keeping with the broader cultural shift towards individual psychological welfare and development after the psychological damage wrought by trench warfare.[53] The influential Hawthorne Factory experiments demonstrated, for the first time, the extent to which scrutiny and interest positively altered staff performance.[54] This was a culture focused on business development as related to individual mental wellbeing. After World War II, however, the landscape changed for organisational management. This change is often attributed to the invention of complex logistics during the war; we might broaden this narrative to consider the other mechanical and social systems approaches that emerged, all tied to technological development. Encouraging staff contentment and productivity became a secondary concern; people, after all, were simply parts of larger systems of control.

This changing approach to organisational management was a new tactic in increasing staff productivity. It also reflected the growing importance of the machine for all forms of production. This was to be a change that would progressively alter working conditions over the course of the twentieth century, something that was anticipated by Norbert Wiener. In a statement that places the machine firmly within the economics of production, Wiener reflected: 'Let us remember that the automatic machine is the precise economic equivalent of slave labor. Any labor which competes with slave labor must accept the economic consequences of slave labor.'[55] Wiener anticipated that the process of mechanisation taking place around him would gradually lead to the replacement of human labour with mechanical 'slaves', and that this change brought with it a new economy with both gains and losses. Many forms of employment would cease to exist. Those who stood to benefit economically from mechanical labour were no longer concerned solely with their workforce but also with machines; machines that could complete tasks without human intervention and which were, in this regard, wage-free labourers or slaves.

In invoking the spectre of slave labour, Wiener referenced the ingrained issue of power and control upon which ground cybernetics was born: that of the servomechanism or 'slavemotor' with the human governor.[56] If we recall that pedagogy is also etymologically rooted in slave labour and education as a controlled process, then it is clear that the interlaced issues of human and mechanical control in post-war behaviourism opened critical questions of power and control in art education as a system of production. The entwined relationship of man, mechanism and system of production meant that the mechanical 'slave labour' and the human paid labour had to be managed holistically – treated as part of the same system of control. The systems relationship described here anticipated Actor Network Theory (ANT) in which both people and *things* have agency within a network. Bruno Latour wrote:

> Students of technology are never faced with people on the one hand and things on the other, they are faced with programs of action, sections of which are endowed to parts of humans, while other sections are entrusted to parts of nonhumans.[57]

The place of ANT within sociological theory echoes the prophetic vision of first wave cybernetics, as scientists explored the philosophical and practical implications of a networked age of technology in which machines had behaviour. Our wired-in world is forged from the complex and wonderful convergences of technology and the human mind. As we trace this development back to its genesis, we must consider the changing power relationships at play as a result of the 'slave labour' provided by machines. We must understand both the evolving need, in the post-war years, to treat technology as a productive work force to be managed and how this related to the management of people within larger frameworks of mechanical–human production. In this context, control derived from an ability to manage connected components, both human and technological.

Groundcourse and the cybernetics of education

Ascott took an innovative approach to teaching from his earliest employment. This began with Groundcourse, Ascott's two-year foundation course, which took place first at Ealing Art College and then at Ipswich Art School in the period 1961–5. It drew in cybernetics, behavioural psychology and analogue computing. While there was certainly a burst of experimental teaching in the immediate post-war period, Groundcourse had a sustained ideology that made it one of the most experimental teaching models of the twentieth century. It was amongst the UK's first art and design foundation courses and it incorporated and interrogated wartime technologies within a framework of conceptualised visual arts projects. Ascott's early practice and his first pedagogical model offered some of the earliest developments in systems and cybernetic approaches to art production. Because of this, Ascott passed on cybernetic principles to students over the course of the early 1960s. It is worth noting that his open, networked approach to art teaching meant that ideas were shared between staff and students with input from frequent visiting speakers. The course evolved as it progressed. Students went on to diverse careers in areas including arts, marketing, music, light design and advertising, and, famously, each incarnation of Groundcourse influenced a number of future musicians, including Brian Eno and Pete Townshend. What is notable from talking to former students is that Groundcourse had an enduring influence for them and that many of their number continued to work in a collaborative way, and also maintained an interest in feedback, communication and cybernetics.[58]

For Ascott, the course was an organism: an entity within which staff and students played parts in its function. The organism was also a frequent point of reference in cybernetic discourse. Ashby, for example, wrote extensively about the symbiotic relationship an organism had with its wider environment: 'Given an organism, its environment is defined as those variables whose changes most affect the organism, and those variables which are changed by the organism's behaviour.'[59] Essentially, Ashby argued that the function of an organism had to be understood in terms of its relationship to its wider environment; it takes in and gives out in a continuous process. Within the field of cybernetics, the organism was a theoretical model: a living system that might be improved or replicated through technology. The word *cyborg* is itself a contraction of the term *cybernetic organism*.

By defining Groundcourse as an organism, Ascott meant that it was not to be defined by its boundaries as an art foundation course. Instead, it was to be a system in continuous exchange with its wider environment. In this cross-disciplinary field, Groundcourse was a pedagogical design, a formal pre-diploma qualification, a collective, a power structure, a social organism, a mechanism and an experiment. It was also a work of art in its totality: created by numerous producers who were themselves part of a larger machine of production. It is precisely this complexity that made Groundcourse a system. Given the collective nature of art training, pedagogy was the perfect level on which to explore systems of production, enacting the dynamics between individuals in this new, networked age for culture.

The course also introduced an unprecedented focus on interactivity to the art school studio. Bearing in mind that participants were as young as sixteen and were not yet even studying for their Diploma in Art and Design (DipAD), this conceptualism is all the more remarkable.[60] Furthermore, leading the foundation course at Ealing was Ascott's own first teaching appointment and as such represents a remarkable achievement in the realm of visual arts pedagogy. Groundcourse incorporated elements of Ascott's own Basic Design training in Newcastle at the King's College campus of Durham University, while also evolving towards this extraordinary interrogation of contemporaneous cybernetic theories. The curriculum was entirely conceived as a system, drawing in cybernetics, analogue computing and behavioural psychology.

In *Design for a Brain*, a book that had considerable potency for Ascott, W. Ross Ashby wrote: 'The organism affects the environment, and the environment affects the organism: such a system is said to have "feedback".'[61] The systems approach to mechanism can be extended outwards to include any number of environmental or outside factors that have the power to affect its function. Ascott's pedagogy clearly demonstrated the concept of feedback; the evocative use of technologies (and psychological approaches derived from technologies) to inform art teaching treated the visual arts as a system of interaction. I argue that Ascott's pedagogy marked a broader change within schools of art in which subjective modes of communication disrupted the materially driven disciplines of the school of art. In return, the application, exploration and subversion of technologies within art education 'fed back' into culture as this generation of artists left education and developed a more subjective, networked mode of practice than any other age.

Thesaurus

In 1962, Ascott produced a work of sculpture called *Video Roget*. It was exhibited at the Molton Gallery in London the following year and Ascott designed a catalogue that featured a reproduction of *Video Roget* overlaid with text printed on tracing paper that contained a lexicon for the sculpture, from art strategies to calibration. Ascott's reference to *Roget's Thesaurus* in the title related to the symbols the work contained, each of which had multiple meanings depending on context. In the same catalogue, Ascott had designed a diagram that was essentially intended as a manifesto for his cybernetic art practice.[62] In fact, it is a manifesto that is broadly applicable to cybernetic art in all its incarnations in the period 1950–70. The terms and concepts that Ascott developed for this exhibition formed the basis of the next decade of his career and in light of this, each chapter of this book focuses on one of these key terms.

Charting Ascott's early career, each chapter unfolds one of his concepts in relation to cybernetic art. The book takes a decentred approach to documenting Ascott's approaches to art and pedagogy, setting his achievements in the context of British art of the period and setting his concepts against contemporaneous debates in art and design. As a cybernetic artist, Ascott recognises that the artist works within a system of influences, receiving and transmitting information and ideas that transform along the way. In light of this the book draws in the early influences on his own career and the many individuals influenced in turn by Ascott, and it traces the origin and transformation of cybernetic ideas, from the material to the conceptual. The first three chapters of the book deal with Ascott's early career and the second three chapters focus on Groundcourse and the core issues of collaboration, control and play. The concluding chapter follows the dispersed influence of Groundcourse by exploring a number of countercultural multimedia projects that took place in the late 1960s in Britain, many of which involved former students.

The first chapter, *Metaform* traces the overlap between technology and biology in post-war British art, particularly the debates and approaches that formed the curriculum at King's College fine art department during Ascott's degree. I question both the genesis and reach of early cybernetic art in the UK, looking at the relationship between the mechanical biology that inspired late abstract artists and art pedagogy following World War II.

Looking first at debates around biology, technology, process and form in relation to art and pedagogy of Victor Pasmore and Richard Hamilton, I then chart Ascott's early development during and immediately after his studies, when he developed an interest in biological metaphors for process and, importantly, in interactivity. I outline Ascott's concept of the 'meta-form' and explore interactive abstract art as a prelude to conceptual art.

Describing Ascott's first exposure to cybernetic theory, 'Analogue' then looks closely at the knot of ideas surrounding his *Analogues,* a series of interactive sculptures he created in the early 1960s that had moveable elements. The term 'analogue' evolved from the Greek *analogos*, meaning proportionate. To use the term 'analogue' for art objects was to create a conceptualised or symbolic meaning; each sculptural object analogous with a meaning, idea or experience. The technologies of World War II – including radars, gun control and missiles – used analogue computers. Looking at the work produced for two exhibitions in the early 1960s, the chapter presents Ascott's 'manifesto' for cybernetic art as well as considering how sculptural form and function evolved from the kinetic to the analogue, with reference to cybernetics and to war technologies.

The chapter 'Field' then addresses the interrelationship of field theories and the aerial view, placing Ascott's early work in the context of countercultural interests in mapping, charting and 'reading' the landscape, as well as exploring his subsequent interest in tables and the horizontal plane of human communication. As a young man growing up in the vicinity of Glastonbury, Ascott developed an awareness of the beginnings of the 'Earth Mysteries' movement, which brought together ancient monuments and science fictions speculation in searching for pattern and meaning in the landscape. Taken in combination with Ascott's vital years in Fighter Control, these more spiritual approaches to the aerial view fuse in his early work with the communication fields produced by contemporaneous technologies. The aerial photograph, the map and field theories of consciousness together express the multiple, and often surprising, interactions between technological development and cosmic speculation that informed the British countercultural movement.

'Control' is the first chapter dedicated to Groundcourse and it unravels the convergences of cybernetics, biology and behaviourism during World War II, arguing that bio-technical models of control filtered into the visual arts in the post-war years, something particularly evident in art pedagogy including Ascott's Groundcourse. I argue that the fact the majority of British artist-teachers had experienced either military service or national service had a discernible influence upon the development of radical art pedagogies. The following chapter "Calibrator" charts Ascott's curriculum, demonstrating how it sought to disrupt the concept of art education which students had experienced at school and to strip back any preconceptions about the subject. Much like their Basic Design predecessors, Ascott and his staff used surprises and disruptions to achieve this, amongst other tactics. However,

Ascott's model had a much more developed focus upon behaviour, and the first part of Groundcourse training was designed to make students reassess their own habits and responses with the aim of making them more open to possibility. This chapter looks at these shock tactics and exercises in self-awareness as cybernetic – and behavioural – approaches to pedagogy.

I then outline the second phase of Ascott's pedagogical model in which students were treated as parts of a greater system: an organism. At the same time, the student body was subject to teaching experiments in psychological conditioning and objective control which saw staff scrutinise the actions and reactions of students in created circumstances. Set against the paranoia of the Cold War environment, the science of predicting and controlling human behaviour was one of the defining ideologies of the era. The chapter examines the politics of pedagogy with regards to behaviourism.

The chapter 'Game' explores play as an essential part of the Groundcourse pedagogy. I form visual comparisons between the objects and environments of the bunker and these art school games. Using photographic records, I demonstrate how student games drew together the elements of behaviourism, analogue mechanisms, group dynamics and interactivity that characterised the course and how these elements are still key to game design. I show how the relationship of the analogue sculpture to the game was played out within the context of Groundcourse – the systemic or mechanistic elements of game design emphasised. Furthermore, in the symbolism and dramatic play of the games constructed by students, there is evidence of the impact of World War II and the Cold War. Within the course pedagogy, games served as exercises in systemic thinking, in codified design, in environmental awareness and, interestingly, as prompts for behavioural analysis. The games design activities provoked a kind of self-conscious tension in participants; the knowledge of surveillance appeared to provoke a weighty, concentrated form of 'play'. In the creative outcomes of 'performing' the games under conditions of observation, there are rich and surprising parallels with the secret and pressurised environment of fighter control.

The book concludes with 'Synthesizer', which traces cybernetic ideas from Groundcourse to British countercultural happenings, particularly rock shows and the art and design that made them spectacular, hypnotic events. Both the Basic Design movement and Groundcourse had amongst their alumni some of the most influential rock stars of their generation. Brian Eno and Pete Townshend were Groundcourse students and, like Ascott, Bryan Ferry had trained at King's College, under the tutelage of Richard Hamilton and Victor Pasmore. While the importance of their training has been long acknowledged, the specific links between Ascott's conceptual use of the analogue and the subsequent application of analogue synthesisers on the part of musician–artist alumni holds arresting possibilities for the dissemination of early electronic music, as does the lightshow phenomenon that was so intrinsic to bands such as Pink Floyd and Soft Machine. This chapter delves deeper into the cybernetic roots of these bands, probing both the material and conceptual application of the cybernetic term ‘feedback’ to creative practice. I analyse the rich and surprising links between cybernetics and both sound and light, as well as the fact that the birth of multimedia was influenced not only by the newly available synthesisers but also by cybernetic theory as a model of creative practice.

The objects and physical environments which were created or explored within Ascott’s pedagogy derived principally from the machines, systems and mechanisms of war. These objects formed a continuous and evolving legacy of the war within visual culture; they were visible, tangible and increasingly symbolic. Computers and the plethora of associated technological forms developed during and after the war have striking and remarkable aesthetic qualities. These new technologies were connected, networked to each other and to their users, which caused new technological environments to emerge. These environments in turn had a strong visual identity. This development – from single mechanical object to networked environment – has a marked symmetry with simultaneous developments in the visual arts from object to concept, from isolated practice to broadly networked, subjective and interactive modes of practice.

Notes

1 Roy Ascott (2015) In conversation with the author.
2 Ascott began to read and respond to cybernetic theory from 1959, while he was a studio demonstrator at King’s College, the University of Durham in Newcastle.
3 Ascott is the founding president of the Planetary Collegium and director of the Centre for Advanced Inquiry in Integrative Arts at the University of Plymouth. He is De Tao Master of Technoetic Arts at Beijing De Tao Masters Academy, Shanghai. He teaches, lectures and exhibits globally.
4 Charlie Gere (2002) *Digital Culture*. London: Reaktion. 373.
5 Edward Shanken’s thoughtful overview of Ascott’s career is the exception: In: Roy Ascott & Edward Shanken (2003) ‘Introduction’. *Telematic Embrace: Visionary Theories of Art, Technology and Consciousness*. Berkeley: University of California Press.
6 I borrow this term from Tim Stott, who used it in discussion at the ‘Mapping Cybernetic Art’ Conference session, Association of Art Historians Annual Conference 2016. This session was convened by the author, Dawna Schuld and Jon Wood.
7 While there were numerous famous collaborations such as Experiments in Art and Technology and the Bell Laboratories residencies, the more informal exchanges on both sides of the Atlantic are equally important. Conferences, visiting lectures, friendships and smaller collaborations abounded. A selection of those concerning Ascott will be detailed in these pages.
8 Herbert Read (1943) *Education Through Art*. London: Faber and Faber.
9 Herbert Read (1966) *The Redemption of the Robot*. New York: Trident Press. 54
10 Herbert Read (1955) *The Grassroots of Art*. Revised Edition. New York: Wittenborn. 108.
11 For an account of teaching at King’s College see: Richard Yeomans (1988) ‘Basic Design and Richard Hamilton’s Teaching’. *The Journal of Art and Design Education*, Vol. 7. No. 2. 155–6.
12 W. Ross Ashby & R. Valée (1951) ‘Statistical Machinery’. *Thalès*, Vol. 7, 1–8, 1.

13 Frederic Jameson (1991) *Postmodernism, Or the Cultural Logic of Late Capitalism*. Durham: Duke University Press. 37.
14 For an account of the workings of Ashby's homeostat, see Hallvard Haug (2016) 'The Thinking Machine: W. Ross Ashby and the Homeostat'. Available online at http://blogs.bl.uk/science/2016/04/the-thinking-machine.html. Accessed 10/10/17.
15 Norbert Weiner (1956) *The Human Use of Human Beings: Cybernetics and Society*. Anchor Books. New York. 95–6.
16 *Ibid.*, 1.
17 *Ibid.*, 6.
18 *Ibid.*, 7.
19 W. Ross Ashby. (1960). *Design for a Brain*. London: Chapman and Hall. 383
20 Claude E. Shannon (1948) 'A Mathematical Theory of Communciation'. *Bell System Technical Journal*, Vol. 27. No. 3, 379–423.
21 Jon Bird & Ezequiel Di Paolo 'Gordon Pask: His Maverick Machines'. In: Philip Husbands, Owen Holland & Michael Wheeler (2008) *The Mechanical Mind in History*. 187–209, 201.
22 Peter Cariani (1993) 'To Evolve an Ear: Epistemological Implications of Gordon Pask's Electrochemical Devices' *Systems Research*, Vol. 10. No. 3, 19–33, 22.
23 *Ibid.*, 22.
24 Click here to enter text. Bird & Di Paolo (2008) 201.
25 *Ibid.*, 201.
26 *Ibid.*, 201.
27 Stafford Beer (2001) 'A Filigree Friendship'. *Kybernetes*, Vol. 30. No. 5/6, 551–9, 555.
28 Gordon Pask quoted in Beer (2001) 555.
29 Bird & Di Paolo (2008) 21.
30 Ashby (1957) 88. Also see 'Analogue' chapter for further discussion of the black box problem.
31 Beer (2001) 555.
32 Kate Sloan *Interview with Roy Ascott* (2011).
33 *Ibid.*
34 Roy Ascott & Kate Sloan in conversation. March 2017. Henry Moore Foundation, Leeds.
35 Shannon (1948).
36 Norbert Wiener (1950) *The Human Use of Human Beings*. Boston: Da Capo Press, 16.
37 Edward Shanken (2002) 'Cybernetics and Art: Cultural Convergences in the 1960s'. In: Bruce Clarke & Linda Dalrymple Henderson (eds) *From Energy to Information: Representation in Science and Technology, Art, and Literature*. Stanford: Stanford University Press. 155–77.
38 See: Jasia Reichardt (1968) *Press Release for Cybernetic Serendipity*. Available online at: http://cyberneticserendipity.net/, accessed 01/04/2013, and Shanken (2002) 'Cybernetics and Art: Cultural Convergences in the 1960s'.
39 Ross William Ashby (1960) *Design for a Brain*. London: Chapman and Hall.
40 The Hawthorne Report from the interwar period remained influential after World War II. It was based on a series of experiments on workers in a factory called the Hawthorne Works, which proved that workers' behaviour modified when an interest was taken in them. See: Gillespie, R. (1991) *Manufacturing Knowledge: A History of the Hawthorne Experiments*. Cambridge: Cambridge University Press.
41 See: B.F. Skinner (1953) *Science and Human Behaviour*. London: Pearson Education Inc.
42 Roy Ascott (1968) 'Cybernetics – Letter to the editor'. *Studio International*, Vol. 176. No. 902, 8.

43 Usselmann, R. (2003) 'The Dilemma of Media Art: Cybernetic Serendipity at the ICA London'. *Leonardo*, Vol. 36. No. 5, 389–6.
44 See B.F. Skinner (1938) *The Behavior of Organisms: An Experimental Analysis*. New York: D. Appleton & Co; (1953) *Science & Human Behavior*. New York: Simon & Schuster; and (1948) *Walden Two*. Indianapolis: Hackett Publishing
45 James H. Capshew (1993) 'Engineering Behavior: Project Pigeon, World War II, and the Conditioning of B.F. Skinner' *Technology and Culture*, Vol. 34. No. 4, Special Issue: Biomedical and Behavioral Technology (Oct.), 835–57.
46 *Ibid.*
47 *Ibid.*, 835.
48 *Ibid.*, 836.
49 N. Katherine Hayles (1994) 'Boundary Disputes: Homeostasis, Reflexivity, and the Foundations of Cybernetics'. *Configurations*, Vol. 2. No. 3, 441–67, 457.
50 *Ibid.*, 457.
51 *Ibid.*, 457.
52 Norbert Wiener (1988) [1950] *The Human Use of Human Beings: Cybernetics and Society*. Boston: Da Capo Press.
53 The Hawthorne Report was widely influential at the time, based on a series of experiments on workers in a factory called the Hawthorne Works which had proved that workers' behaviour modified when an interest was taken in them. See Gillespie (1991) *Manufacturing Knowledge: A History of the Hawthorne Experiments*. Cambridge: Cambridge University Press.
54 *Ibid.*
55 Norbert Wiener (1989) [1950] *The Human Use of Human Beings*. London: Free Association Books. 162.
56 Norbert Wiener (1948) *Cybernetics, or Control and Communication in the Animal and the Machine*. Massachusetts: MIT Press.
57 Bruno Latour 'Where are the Missing Masses: The Sociology of a Few Mundane Artifacts'. In: Wiebe E. Bijker & John Law (eds) (1992) *Shaping Technology: Building Society Studies in Sociotechnical Change*. Cambridge: MIT Press.
58 I interviewed a number of Groundcourse students between 2009 and 2018 and this interview material is used throughout the book.
59 R.W. Ashby (1960) *Design for a Brain*. London: Chapman and Hall, 36.
60 In 1960, the new Diploma in Art and Design was introduced, which was intended to have degree equivalence. See: Coldstream, W. (1960) *The First Report of the National Advisory Council on Art Education*.
61 Ashby, R.W. (1960) *Design for a Brain*. London: Chapman and Hall, 37.
62 Roy Ascott, *Diagram Boxes and Analogue Structures*. London: Molton Gallery.

1 Metaform

Biology | mechanics | structure

Form, growth and process

In Britain after World War II, artistic engagement with biology took an increasingly mechanical turn. After a war that had seen British biologists and psychiatrists commissioned to engineering, code-breaking and logistics, a more holistic view of systems behaviour, from the organismic to the technological, gradually infiltrated every discipline.[1] D'Arcy Wentworth Thompson's *On Growth and Form* remained a powerful influence for British artists, although its central thesis of treating biological form as process was also gaining new relevance and complexity in relation to theories of field and environment.[2] In 1951, debates about the new biology were given a platform when Richard Hamilton's exhibition *Growth and Form* and the accompanying symposium *Aspects of Form* took place at the Institute of Contemporary Arts in London, during the Festival of Britain. Also in the period 1945–55, several artists who were to contribute to the Basic Design movement in British art education were working at schools of art across London, but principally at the Central School of Arts and Crafts.[3] This chapter will chart the debates about biological form and new technologies that preceded and informed the development of the Basic Design movement, outlining how the same ideas were incorporated into the art and design curriculum at King's College. It will connect the manifestations of biological form-as-process in late British abstract art to the development of interactivity in the work of Ascott and his contemporaries.

In the immediate post-war years several British artists engaged with the mechanics of form, some of their number connecting their interest in form to cybernetics and the enclosing question of systems behaviour. Amongst their number were both Hamilton and Pasmore, who taught Roy Ascott at King's College, the University of Durham in Newcastle from 1955 to 1959.[4] The two artists might have had similar interests, but they often had opposing viewpoints; Hamilton was resolutely future-facing whilst Pasmore remained committed to abstraction for his whole career. Pasmore was at this time a well-respected and active artist, heading the Art Department, contributing to the development of national policy in art education, exhibiting and completing numerous commissions. Beyond this, he held great influence for his pupils, many of whom credit him as the most important figure in their education.[5] Because Pasmore believed in intuitive process, his work was latterly and persistently misunderstood as lacking theoretical grounding.[6] The two most contentious elements of Pasmore's career were his conversion to abstraction from his previous delicate, impressionistic realism, and then, after a spell working with construction, his return to abstract painting at the precise moment the majority of artists had abandoned it. One of the tasks of this chapter is to pursue the philosophy of abstract form that Pasmore taught at King's College, while considering the nature and extent of his influence upon Ascott and his contemporaries. While Hamilton has achieved greater fame and attracted far more research interest than Pasmore, his teaching at King's College has not received the critical attention it deserves.[7] Certainly, Hamilton's pedagogy in the 1950s and 1960s was an extension of his practice, touching upon many of the same themes and ideas that he presented in his art and exhibitions of the same period. It is also evident that Hamilton's consuming interest in – or even idolatry of – machines formed a discernibly influential element of the Basic Design curriculum. For this reason, this chapter seeks out the various points of intersection between biology, engineering and art pedagogy that drove students towards experiments in increasingly systematized and interactive abstract art. Moving on to Ascott's training and early career, I then establish how this training contributed to Ascott's conviction that 'form has behaviour', an idea that has endured in his practice for several decades.

On Growth and Form and the structure-process debate

While Hamilton was still a student, he developed an interest in both biological and engineered form and process. At the Slade, he met Nigel Henderson who first introduced him to *On Growth and Form*. The book quickly became an obsession and occupied much of his remaining time at the Slade. The *Reaper* prints were made directly prior to Hamilton's larger project – the exhibition *Growth and Form* which he proposed and developed for the ICA. The exhibition explored the underlying philosophy of biological structuralism which Thompson had offered as an alternative to the prevalent, and in his view limiting, interest in evolution. It brought together cutting-edge imagery such as photomicrographs as well as scientific models and films, alongside abstract art with relevance to biological structure. None of this work was labelled, to allow for meditation upon the visual and formal qualities of the selected exhibits. In his initial proposal to Herbert Read in 1949, Hamilton pitched his ideas:

> The initial stimulus for the proposed exhibition was provided by Thompson's book On Growth and Form. The visual interest of this field, where biology, chemistry, physics and mathematics overlap was

> considered an excellent subject for presentation in purely visual terms. The laws of growth and form pertaining to the processes of nature are quite contrary to the processes of artistic creation. However complex the form (accepting Thompson's hypothesis) it is the result of very precise physical laws; the complexities of art, on the other hand, are the products of involved psychological processes.[8]

This tentative description of the overlap of arts and sciences is an early formulation of Hamilton's engagement with the processes described in *On Growth and Form*. In comparing the 'involved psychological processes' of art to the 'precise physical laws' of form, Hamilton linked the two with Thompson's structured morphology.

Structure (Figure 1.1) was one of a series of prints Hamilton made at the Slade which explored the concepts he had gathered from D'Arcy Wentworth Thompson. He experimented with repetition of marks to create structured form and composition, relating to growth patterns in the natural world. The outcome has a network quality, the suggestion of lines connecting the decisive, incision-like crosses on the paper – it is a map of the series of decisions that led to the finished abstract image. This print gives a useful insight into

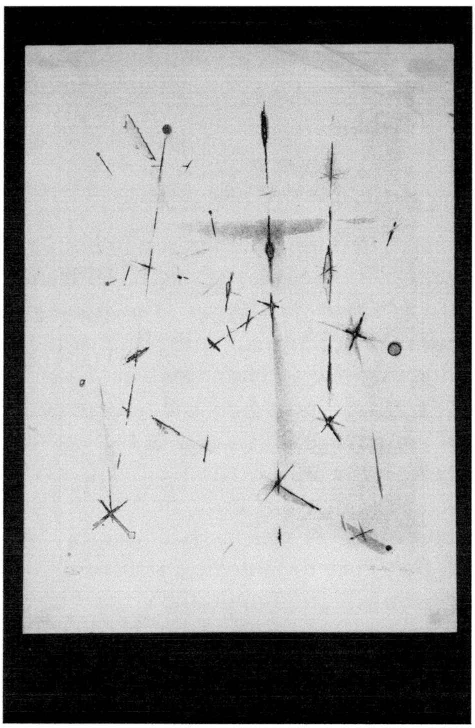

Figure 1.1 Richard Hamilton (1950) *Structure*. Liftground etching and aquatint on paper. 15 3/4 × 11 7/8 Tate Gallery

Hamilton's reductive approach to abstracting Thompson's descriptions of growth patterns, each point of the composition darkly incised to highlight the underlying geometry of natural growth.

When Hamilton developed his ideas about *On Growth and Form* into a formal proposal, he articulated a rational structure for the exhibition, the subject headings of which were:

1 Time as a dimension of form
2 Forms of cells
3 Cell groupings
4 Skeleton structure
5 Related forms
6 Form and mechanical efficiency
7 The formal realization of pure mathematics[9]

This formed the basis for the ICA exhibition, and it offers an interesting illustration of the multidisciplinary approach upon which the ICA was conceived. It is not only the arts and sciences collaboration that made this exhibition an important event, but also that all the scientific disciplines are represented together, exploring the same themes of mathematical structure and growth in the natural world. Note that this linked set of subjects includes 'form and mechanical efficiency', a subject which relates both to Thompson's entreaty to refer all form 'to mechanism' and also Sigfried Giedion's *Mechanization Takes Command*, in which he discusses how biological science had been altered by the mechanization of society.[10] Hamilton himself made this link with Giedion explicit, when in his initial proposal for the ICA exhibition he wrote:

> The most obvious benefits of the exhibition would be the influence it may have upon design trends. The general implications are very wide: S. Giedion in his study of mechanization says 'The evolution from material and mechanistic conceptions must start from a new insight into the nature of matter and organisms'. The exhibition should also make its contribution in this direction.[11]

In the few existing installation shots of *Growth and Form* from 1951 (Figure 1.2), a soft-edged geometry of display is apparent.[12] The abstract screen structure in the foreground is amorphous – reminiscent of cell structures, of bones and of rock formations. It was created by Hamilton as part of his exhibition design. Behind it, there are the open cubic frames that Hamilton used to display the objects, and on the walls, abstract painted forms with the same amorphous quality of the screen – like bones, rocks or drops of water. This fusion of interdisciplinary objects and abstract imagery was a strategy Hamilton would employ again in his later exhibitions for the ICA, including *Man, Machine and Motion* in 1955.

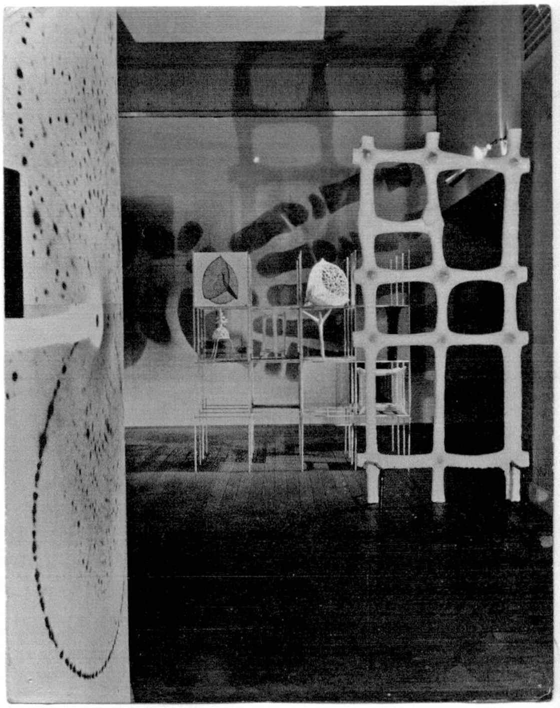

Figure 1.2 Installation view of Richard Hamilton's exhibition *Growth and Form* (1951) ICA archive, Tate Gallery

The cubic display structures employed by Hamilton have as much interest as the exhibits themselves. Hamilton commented that:

> Growth and Form seemed an ideal subject for another involvement of that time, exhibition design. By the turn of the century the 'exhibition' was beginning to be understood as a form in its own right with unique properties. My meeting with Roland Penrose was propitious because he commended the idea of an exhibition on Growth and Form to the Institute of Contemporary Arts. The result was that a good deal of time spent at the Slade was devoted to finding the financial resources, researching, designing and, in part, making the exhibition which was to be the ICA's contribution to the Festival of Britain in 1951.[13]

The exhibition as a creative undertaking was, as Hamilton noted, a twentieth-century notion. Hamilton's use of geometric frameworks emphasized both his interest in the mathematics of form and also the related issue of repetitions and groupings. It is notable that here the white cubic geometries which so dominate display aesthetics in contemporary art have a subtext of reason and order. His grid-structures were both constructions of cells and single forms, networks of repeated shapes, and also hint at the measured proportions of natural structures within Thompson's book. The publicity material for the exhibition (Figure 1.3) was based around a Cartesian grid too, the visuals and text displayed in an off-centre arrangement. The images include cells, crystals and the beautiful hexagonal skeletal structure of the Radiolaria, as discussed by Thompson himself at length.

Thompson wrote that scientists needed to pursue 'the ephemeral and accidental, not eternal nor universal things' in order to gain sophisticated understanding of the physical world.[14] This is a striking echo of Charles Baudelaire's famous call that artists should capture modernity: 'By "modernity" I mean the ephemeral, the fugitive, the contingent, the half of art whose other half is the eternal and immutable [...]'[15] This is a rich parallel between modernity in art and modernity in science, representing as it does the same collision of an established value system based on a teleology with the scrutiny of nature (and humanity) as it unfolds around us. For Thompson, natural forms were growing, evolving structures, linked into a web of potential influences and outcomes. Thompson's biology treated the organism as part of an extended environment, another indication of how his contribution to both biology and art converged with early systems thinking. One could also argue that the growing interest in the contexts of display and interpretation on the part of artists such as Hamilton and Pasmore was essentially a systems approach, drawing in as they did not only the abstract work of art but its environment, its audiences and its relationship with other disciplines.

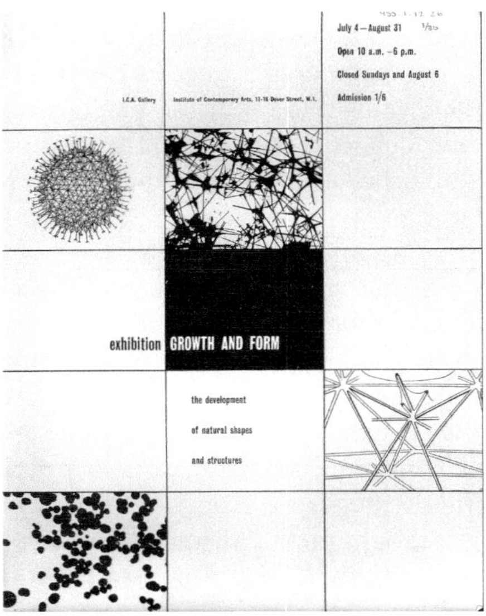

Figure 1.3 Publicity material from the *Growth and Form* exhibition (1951) Tate Gallery ICA Archive. Ref. TGA. 955/1/12/26

On Growth and Form and the propositions Thompson offered about the mathematics of biological form proved to be an invaluable source of ideas for artists and designers in the twentieth century. The importance of biological form to British modernism (such as the influence of morphology, evolutionary transitions, organic form or process) and the place of *On Growth and Form* in this dialogue is well-documented.[16] There is another layer to this relationship which has been neglected: this is the *mechanization* of biological form and its subsequent influence on abstract art. Thompson writes:

> In Aristotle's parable, the house is there that men may live in it; but it is also there because the builders have laid one stone upon another. It is as a *mechanism,* or a mechanical construction, that the physicist looks upon the world, and Democritus, first of all physicists and one of the greatest of the Greeks, chose to refer all natural phenomena to mechanism and set the final cause aside.[17]

In pursuing the factors (other than evolution) which dictate growth patterns and eventual structure in natural forms, Thompson necessarily treated biological form as mechanism. Thompson's position is as rooted in philosophy as it is in practice. As well as attempting to overcome biology's dominant focus on evolution in favour of a broader, mathematically oriented approach, he also makes the above point in opposition to the teleological problem for scientists – that of the world as God's creation. In this Aristotelian parable, even if the world was created for man it was also *built*, brick by brick. Thompson thus argues that biologists must look at process in order to understand the final form, rather than the underlying cause. Thus *mechanism* in this context is the processes and influences that contribute to the evolving structure of a living form. This in itself is in contrast to the methodological reductionism which had dominated the sciences for the previous two centuries, based on giving research problems or areas fixed and defined boundaries. The cost of this is the exclusion of elements which influence each other within the complex interactions of systems of knowledge.

To an extent then, Thompson's argument within *On Growth and Form* anticipated the systems theories that were to emerge in the interwar period. This is because when the study of biological growth is opened out to mechanical reasoning, by extension all processes and influences upon that living form must be examined, creating a network or system around the living form. For example, plant growth and its evolving form involve variables such as climate, weather condition, soil, disease and proximity of other plants. Where the reductionist might exclude variables to create boundaries, the mechanical biologist would pursue them. Thompson gives the example of an amoeba as a 'so-called simple organism', stating some of the forces at work and then commenting that:

> Like other fluid bodies, its surface, whatsoever other substance – gas, liquid or solid – it be in contact with, and in varying degree according to

> the nature of that adjacent substance, is the seat of molecular force exhibiting itself as surface-tension, from the action of which many important consequences follow, greatly affecting the form of the fluid surface.[18]

If we were to map the mechanics of a simple organism in terms of forces upon it, what would result would be a connective network of elements: gases, liquids, and solid forms which contribute to its processes, thus extending far beyond the reductive boundaries of traditional morphological analysis. The extent to which the reductionist tradition had dominated science must be emphasized here – in the natural sciences, all living creatures had been treated as closed systems by scientists for the purposes of 'demarcation'. Of pre-twentieth century biological science, Giedion wrote that: '[...] in biology, the animate being was considered simply as a sum of its parts, assembled like those of a machine. Organic processes were regarded as purely physico-chemical in nature, as if an organism were a kind of chemical plant.'[19] He saw the twentieth century – more specifically the interwar period – as the moment of completion for the process of mechanization, and thus the period in which other disciplines, including biology changed their practices due to understanding mechanized interconnectivity.[20] The essential work by the biologist of analyzing physical forces upon form in turn relies upon the most important factor in Thompson's vision: the Platonic notion of ideal numbers. Stephen Jay Gould writes:

> (1) His view of Plato and Pythagoras (mathematics, generality and deduction) versus Aristotle (description and induction). (2) His Greek commitment to a pure and abstract understanding of form versus his Baconian idea that knowledge is power, as expressed by the engineer's love for a good design because it works.[21]

In his 1971 article in *New Literary History,* Gould explores the philosophical construction of Thompson's vision, and its continuing relevance and influence in the twentieth century. Above, he outlines the theoretical basis for *On Growth and Form*, a synthesis of classical mathematics and contemporary methodology. Thompson argued that the pure mathematics that is so essential to the physical sciences was essential also to the natural sciences, citing Bichat, Pascal and Schwann.[22] He proposed that the language of mathematics is a unifying element across the sciences, and in emphasizing this, Thompson once again drew together similar systems to be found within traditionally distinct areas of scientific research.

I make this point not because there is an established trajectory of ideas between D'Arcy Wentworth Thompson and the later systems theorists, but rather because there is a fundamental philosophical shift here which underpins both. This is the move away from reductionism towards a scientific practice which interrogates the complex relationship of influences around any given object of study. Ludwig von Bertalanffy believed that 'So far the unification of science has been seen in the reduction of all science to physics, the final resolution of all sciences to physical events.'[23] This was certainly true of Thompson and *On Growth and Form,* in that he wanted to '[...] correlate with mathematical statement and physical law certain of the simpler outward phenomena of organic growth and structure or form [...]'[24]

That these laws also dictated the outcomes of human endeavours such as architecture and engineering was also evident to Thompson. This is best evidenced by the parallels Thompson formed between engineering and biological structure in his chapter 'On Form and Mechanical Efficiency', in which he compares the engineering of bridges to bone structure, and recounts an anecdote about the engineer Professor Culmann, who upon visiting the dissecting room of his colleague Meyers and viewing a section of bone, reputedly cried out 'That's my crane!' (Figure 1.4).[25]

Thompson explains:

> [...] we have no difficulty in seeing that the anatomical arrangement of the trabeculae follows precisely the mechanical distribution of compressive and tensile stress or, in other words, accords perfectly with the theoretical stress-diagram of the crane.'[26]

The same physical laws underlie both mechanisms, organic and manmade, as both fulfil the same purpose of distributing stress while maintaining form. The manmade world is not set apart from these physical laws: every known form, manmade or otherwise, is formed by function and environment. Any mechanism is, quite simply, a system.

During *Growth and Form,* the ICA held a symposium called 'Aspects of Form'. A book of essays was published under the same name to coincide with the exhibition, and it is extraordinary in the extent to which it suggested early systems thinking. Herbert Read wrote in the preface that: 'The increasing significance given to *form* or *pattern* in various branches of science has suggested the possibility of a certain parallelism, if not identity, in

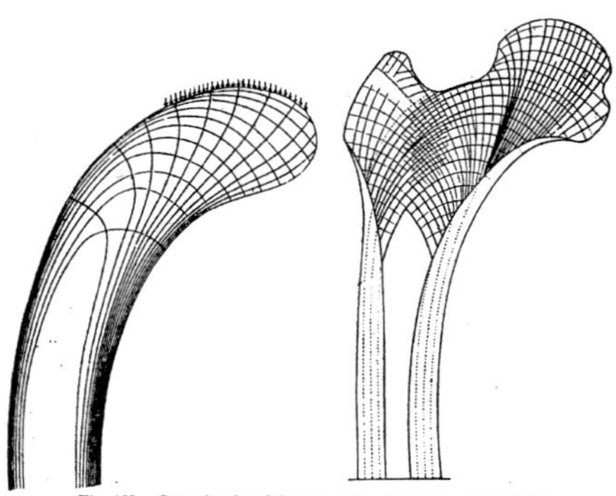

Fig. 463. Crane-head and femur. After Culmann and J. Wolff.

Figure 1.4 Crane-Head and Femur (After Culmann and J. Woolf). From D'Arcy Wentworth Thompson. *On Growth and Form.* 1917, 233

the structures of natural phenomena and of authentic works of art.'[27] The book was edited by Lancelot Law Whyte, who believed that there were common patterns exhibited across disciplines which needed to be explored. While Read commented on the possible parallels between the formal qualities of modern art and those in the natural world, Whyte gave an even more ambitious suggestion in his introduction to the book:

> Common to the ideas of form, configuration, pattern, and structure, is the notion of an ordered complexity, a multiplicity which is governed by some unifying principle. Our theme is thus the *realization of unity of spatial form in the complex processes of physics, biology, psychology, and art.*[28]

Whyte defines the book as a pursuit of form as clearly set in space, so for his purposes he excludes 'musical form, linguistic form, abstract mathematical form, and the forms of thought, of human personality, and of society'.[29] However, by identifying these alongside physics, biology, psychology and art, Whyte demonstrates a connective or systems approach in line with that which von Bertalanffy would formalize and legitimate. He excluded only those formal structures which could not in his opinion be analyzed visually. This emphasis on parallelism also echoes Giedion's multifaceted exploration of mechanics in the modern world, seeking similar concepts across traditionally separate fields. Whyte's phrase 'an ordered complexity' is expressive of the balance between pattern and chance which dictates form in the natural world. In a 1951 review of the book, crystallographer Kathleen Lonsdale stated that:

> Those who seek a unitary principle which correlates all observations and experiences within the range of human perception and understanding will welcome this volume, which is a general survey of visual form, from physics through biology and psychology to art.[30]

The subjects of chapters within *Aspects of Form* included modern embryology, gestalt perception, activity patterns in the brain, biochemical aspects of form and Rudolph Arnheim's famous contribution on 'Gestalt Psychology and Artistic Form'. Its parallel presentation of formal growth and development across disciplines was an indicator that within the world of contemporary art, a new *modus operandi* was developing. Both the formal qualities of abstraction and the thematic content of works of art were networked with parallel developments in the sciences. In the chapter 'Biochemical Aspects of Form and Growth', Joseph Needham wrote that:

> A unified science of life must inevitably seek to know how one level is connected with others. For the body contains organs, the cells nuclei and mitochondria, these structures are built up of colloidal particles which in turn consist of molecules large and small (proteins, carbohydrates,

> fats, steroids, etc.), within which again are the atoms with their different kinds of valences and bonds.[31]

Between the unitary levels of organisms, there are vital connections of exchange – and for Whyte this connective network also includes aspects of culture, society and economy. This unique collection of essays offers an insight into the creative culture out of which Basic Design grew: a culture of interdisciplinary parallels, analytical practice and gestalt form, of mechanism, structure and connectivity.

Basic form in pedagogy

It was Hamilton's move in 1954 from the Central School to King's College, University of Durham that started the Basic Design practice there. He was hired to teach both commercial design and to help students understand its synthesis with the fine arts too. That same year, Victor Pasmore was recruited by Professor Lawrence Gowing to head up the painting course. Hamilton was already in post, running his basic courses in the design department and undertaking some limited teaching within fine art from then until 1966. The two men had worked together at the Central College of Arts and Crafts, under the leadership of Sir William Johnstone. They were in good company, with Eduardo Paolozzi, William Turnbull and Alan Davie amongst their colleagues, at a time when Johnstone was keen to establish a more future-facing manner of training inspired by the Bauhaus.[32] Indeed, it was at Central that the Basic Design movement in British art education began, with Johnstone's attempts to establish 'basic' or shared courses for artists and designers. The movement consisted of overlapping pedagogies of abstraction, taking place at several art schools but led by Pasmore and Hamilton at King's College and Harry Thubron and Tom Hudson at the Leeds College of Art. Drawing on the Bauhaus approaches that had infiltrated the UK during and after the war, Basic Design teaching models offered students of art and design a shared 'basic' education in form, line, colour and structure – described as a 'grammar' of form by several of the artists involved. A basic grammar of forms could connect artists and designers to industry; it could be open, interrelated for construction/assembly, and above all, modern. Pasmore recalled that he was:

> [...] interested in developing a new visual language which was independent of anything in front of you such as a still life, a language made up of basic visual sensations and basic visual forms. So at Newcastle I concentrated on experimenting with the students in creating this new language, which I called basic form, basic sensation.[33]

Pasmore also recalled that he was strongly influenced by Bauhaus ideas, to which he had been introduced by William Johnstone at Central.[34]

There is much evidence within the anecdotal and critical histories of Basic Design that teaching staff treated their classes almost as an extended practice, using the students within a broader agenda of interrogation of abstraction. In relation to this approach at King's College, Yeomans has noted that: 'This drive towards experiment was spearheaded by Victor Pasmore who regarded the art studio as a laboratory where his teaching went hand in hand with his own creative research.'[35] As a former student of Basic Design at King's College, Yeomans experienced first-hand Pasmore's approach to teaching as an extension of his own experimental practice.[36] In the post-war years King's College pedagogy operated around a conflicting language of abstraction. Biological form occupied a prominent place within the curriculum, an unsurprising fact given the interwar growth of organic abstraction. However, after the technological innovations of World War II, a new biology was emerging. This was biology in transition towards synthesis with technology, a process that had fascinated Hamilton from the late 1940s when he first started working around the interface between the natural world and the engineered world.[37] Pasmore, meanwhile, remained interested in a more intuitive mode of abstraction, aligning art-making processes with growth processes in nature.

Despite Hamilton and Pasmore having divergent approaches to both teaching and their own art practices, there were several formal links between what they achieved under the banner of Basic Design. Despite Pasmore's intuitive brand of abstraction, he had many of the same theoretical and formal influences as Hamilton – particularly mechanical biology. The conflicting approaches of the two men created two strains of abstract pedagogy that they would later teach at King's College: one analytical and formal, the other intuitive, aesthetically judged and open. However, underlying Pasmore's intuitive organicism was a systems awareness. In arguing against what he believed was a confusing definition of so-called 'structural processes' he wrote in a letter to the artist Charles Biederman, that:

> [...] the mind is capable of projection as well as reflection. This means that the artist is a creative as well as an imitative being. He is not compelled to copy nature because he himself is a process of nature containing the very same elements which he sets out to copy. In other words he can produce the likeness of nature out of himself. His mind is conditioned as much by its own structural process as by the processes imposed on it from the outside; hence its power of projection.[38]

Pasmore mentions the structural process of the mind, as well as outside forces, much as Thompson believed that natural forms must be understood in the context of the multifarious influences upon them – natural form as mechanism. This concept of form-as-process certainly informed the process-driven exercises in abstraction that were taught at King's College by both Hamilton and Pasmore.

Hamilton's mechanics of form: engineering and nature

In the converging landscape of science, technology and art, it is no surprise that both mechanical developments in the twentieth century and also the cultural and philosophical shift that mechanical technologies created held an important place in the Basic Design curriculum at King's College. Richard Hamilton had first started forging parallels between developments in technological and biological form while a student himself. He recalled that:

> Sigfried Giedion's *Mechanization Takes Command* became a primary source book immediately after its publication in 1948. It was particularly significant for me in that it complemented *On Growth and Form*, which deals with the natural world in just the wide-ranging manner of Giedion's perception of technological form and process.[39]

Drawing on the particular brand of ephemera produced by manufacture and engineering, Giedion included analysis and illustrations of forgotten patents, catalogues and diagrams for manufacturing processes and tools, as well as research equipment from the biological sciences. In salvaging and accounting for these frail documents, Giedion created a record of mechanical infiltration; the machine shaping every aspect of the modern world. *Mechanization Takes Command* was an influential text for Hamilton and other members of the Independent Group, and Hamilton himself recognized its importance to his own oeuvre.[40] However, as the statement above indicates, Hamilton read Giedion in tandem with *On Growth and Form,* directly before he started work on the 1951 Institute of Contemporary Arts exhibition *Growth and Form*. This fusion of technological and biological sources during the period overcomes the enduring and problematic binary forged between the biological and the mechanical, in the same mode of contrast as urban/rural, natural/manmade. In the introduction to their 2011 book *Biocentrism and Modernism*, Oliver A.I. Botar and Isobel Wünsche highlight the ways in which historians of twentieth-century art have reinforced this binary, commenting: 'When not ignoring the interconnections between nature-centric ideology and Modernism, historians were denying it, emphasizing, instead, its anti-natural, so-called "mechanistic" aspects.'[41]

As the current interest in the rich and complex interface between modernism and biology grows, we are gradually moving away from this problematic offset of the mechanical against the biological in the histories of modern painting and sculpture. Mechanics are as relevant to the plant as to the machine, to systems of all kinds, since mechanics in its truest sense is the mathematics of motion or forces. This understanding has somewhat paled because of the broader cultural association between mechanics and 'the machine', an emblem of the transformative technologies of the modern world. This is a reflection of a dichotomy that was concretized in the interwar years – thinking as the realm of science and feeling as the realm of the arts. Giedion's book was written upon the perceived disjuncture that was developing between science and the other disciplines which would be voiced a decade later by Charles Snow in *The Two Cultures and the Scientific Revolution*.[42]

Giedion had a sustained interest in how design and architecture could respond efficiently and harmoniously to the proportions of the human body. He titled his history as 'anonymous' as it focused on the under-sung and often authorless production of mechanical solutions to problems or processes. He explored diverse areas of contemporary mechanics, including biology, architecture and design, writing that: 'For the historian there are no banal things.... Tools and objects are outgrowths of fundamental attitudes to the world.'[43] This point has immediate resonance with Hamilton's practice, both as an artist and as an educator. This was never clearer than in the 'Analytical Drawing' exercises undertaken by Basic Design students under Hamilton's tuition, reflecting the cultural significance of mechanical form, as emphasized in *Mechanization Takes Command*.

Giedion was primarily a cultural historian, and his first book *Space, Time and Architecture* was the basis of the themes further explored in *Mechanization Takes Command*.[44] Both caused a considerable stir in the creative and scientific communities, but not without criticism. After receiving a copy of *Space, Time and Architecture* from the architect Erich Mendelsohn, who was a critic of Giedion's ideas, Einstein responded in a private letter as follows:

> Dear Mr Mendersohn,
> The passage you sent me from the book *Space, Time and Architecture* has inspired the following reply:
>
> It's never hard some new thought to declare
>
> If any nonsense one will dare
>
> But rarely do you find that novel babble
>
> Is at the same time reasonable.
>
> Cordially yours, Albert Einstein.
>
> P.S. It is simply bull without any rational basis.[45]

Arthur Molella reflected upon this terse response in a 2002 article reviewing Giedion's two books. Commenting on the efforts of scientists in the interwar period to guard the boundaries of their discipline, Molella referenced Karl Popper's description of the problem of 'demarcation'.[46] In the crisis of knowledge after World War I, all disciplines went through a process of interrogation, reassessing the epistemological grounding of research in practice. While Einstein and many of his contemporaries resisted relativism in favour of traditional objectivity and rationalist approaches, the war had shaken the discipline beyond measure.

Hamilton became increasingly preoccupied with the overlapping processes of nature and engineering which he had gleaned from Thompson and Giedion during his formative years, writing: 'Agricultural machinery was seen by Giedion to be at a crucial interface, the boundary at which technology meets nature.'[47] In a chapter of *Mechanization Takes Command* that addressed agricultural design and which influenced Hamilton, Giedion discussed Oliver Evans' mechanical grain mill (Figure 1.5).

Evans had taken the mill from design to manufacture in the late eighteenth century and Giedion recounted the story, which involved rivalries between millers and an attempt to steal the idea, resulting in Thomas Jefferson being called in to mediate. Jefferson was not vastly impressed with this technological advance, as Giedion tells it:

> [...] 'The elevator,' he declared, 'is nothing more than the old Persian Wheel of Egypt, and the conveyor is the same thing as the screw of Archimedes.'[48]

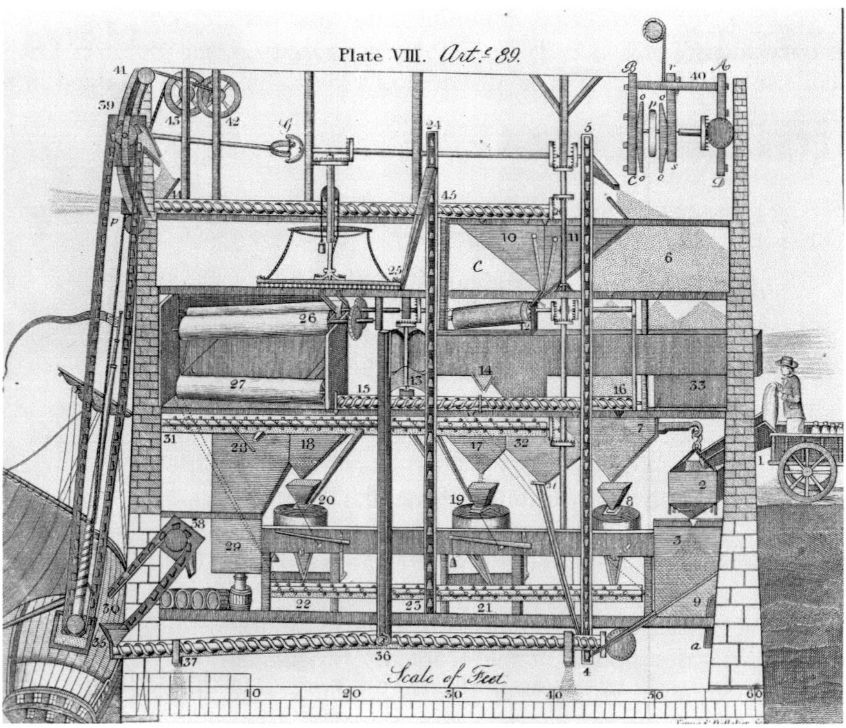

Figure 1.5 Oliver Evans's Mechanised Mill. Reproduced in: Sigfried Giedion (1948) *Mechanization Takes Command: A Contribution to Anonymous History*. Oxford: Oxford University Press

Jefferson connected the mechanical mill with the technological advances in ancient Egypt and Greece upon which it relied, rather than recognizing the machine in its entirety as an innovation. For Hamilton though, agricultural machinery created a 'crucial interface' between nature and technology, which he directly interrogated in his series of Reaper prints of 1949 (Figures 1.6 & 1.7), as he recalls: 'The initial stimulus for a series of twenty Reaper engravings, made at the Slade, undoubtedly came from Giedion's chapter on the farm implement.'[49] In this series of prints Hamilton's reaping machine is reduced to a series of linear and connected lines; a grid that has the clear rationality of a patent diagram.

In *Reaper (e)* (Figure 1.6) the raw teeth of the machine meet the softness of the earth, but the linear relationship of ground to machine is harmonious, suggestive of the beauty of economical mechanical function. With seat, lever and reaping wheel frame reduced to geometric elements, the print emphasizes the logic that underscores mechanical design: form to function, a structure designed for a task and nothing more. Taken in parallel with *On Growth and Form* then, we can draw a comparison between Thompson's analysis of biological growth and Giedion's analysis of mechanical development. Both explore in diverse ways the defining factors of eventual form. Hamilton wrote that: '*On Growth and Form* [...] deals with the natural world in just the wide-ranging manner of Giedion's perception of technological form and process.'[50]

Hamilton identified that the two books by Giedion and Thompson were united by the pursuit of the underlying mechanics of form. The mechanical structures in the *Reaper* prints were both skeletal and diagrammatic: the relationship between technology and nature fused, complete and efficient. Mechanical structure mirrored biological structure because for Hamilton, both had the same rational basis for their eventual form. A patent diagram for McCormick's Virginia Reaper (Figure 1.8), which featured in Giedion's chapter on the mechanization of agriculture, demonstrates the extent of the influence – both visual and philosophical – which Giedion's book had for Hamilton. Their relationship with the *Reaper* print series is striking.[51] The

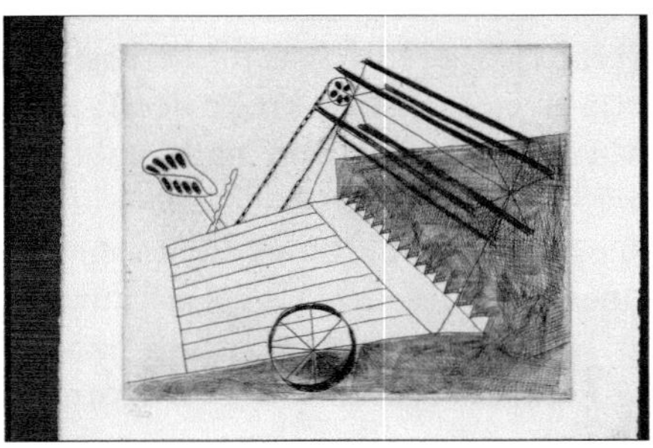

Figure 1.6 Richard Hamilton (1949) *Reaper (e)*. Intaglio print on paper, 175 × 222 mm, Tate Gallery

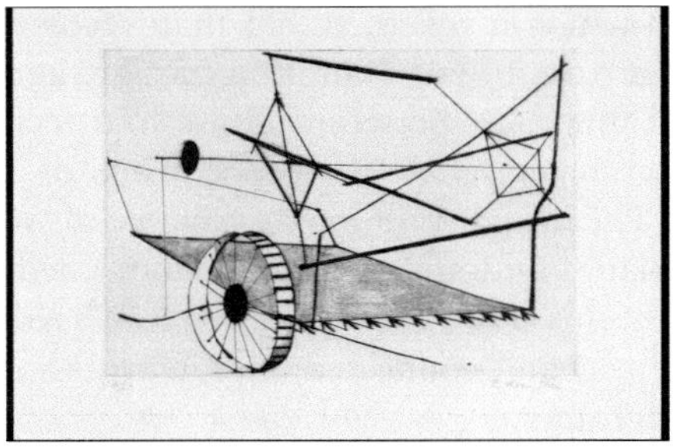

Figure 1.7 Richard Hamilton (1949) *Reaper (h)*. Intaglio print on paper, 171 × 247 mm, Tate Gallery

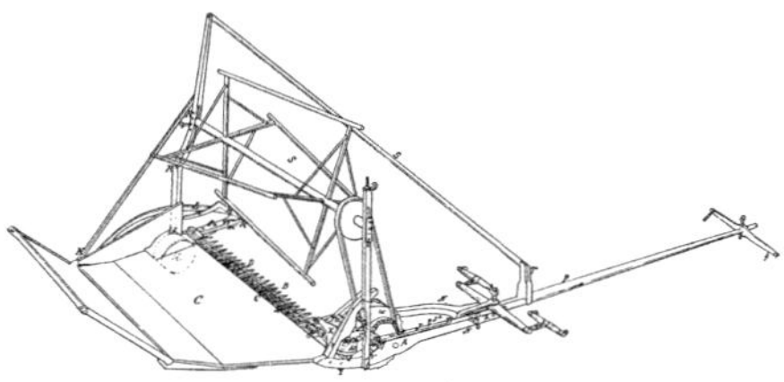

Figure 1.8 *McCormick's First Reaper, Patented January 31, 1845*. Reproduced in Sigfried Giedion (1948) *Mechanization Takes Command*, 135

patent diagram reduces the design to a modernist structure of lines and the woodblock print shows a reaper in use, the machine harmoniously integrated into the farmland around it. Nature and technology meet seamlessly.

Ascott at King's College: from grammar of form to metaform

Basic Design pedagogy was distinctly systematized; it was an attempt to chart the possibilities of abstraction in order that students might have a full comprehension of its possibilities. The concept of a 'grammar' of form appears in writings and commentary by several associated artists of the Basic Design movement, in relation to the creation of courses that broke down a shared, basic visual language. In 1961–2 Maurice de Sausmarez had published two articles entitled 'A Visual Grammar of Form' and 'A Visual Grammar of Form2' in *Motif* journal.[52] In 1959, Richard Hamilton wrote 'Diagrammar', a short article about his teaching methods at King's College, University of Durham.[53] It was published within *The Developing Process*, a book about Basic Design teaching at Leeds and Durham edited by Victor Pasmore and published alongside the ICA exhibition. As well as these prominent articles, there are myriad references to the 'grammar', 'syntax' or 'language' of form in the writing of the key Basic Design figures: Victor Pasmore, Richard Hamilton, William Johnstone, Harry Thubron and Tom Hudson.

The notion of a grammar of form had also been deeply ingrained in the teaching practices at the Bauhaus, best demonstrated by Wassily Kandinsky's *Point and Line to Plane*, writing of the '**art elements** which are the building materials of every work of art'.[54] As Ellen Lupton has commented, his colleague 'Lazlo Moholy-Nagy sought to uncover a rational vocabulary ratified by a shared society and a common humanity'.[55] The Bauhaus pedagogues hoped to create a unifying grammar of form, a basic shared language of human creativity. The Bauhaus ideologies spread across Europe and the United States with the World War II emigration and with the exile of its members. We can certainly read the revival of this grammatical approach in UK schools of art as a Bauhaus inheritance.[56] Hamilton himself recalled that his early teaching for design students at King's College was derived from the Bauhaus ideas he had been teaching at the Central School.[57] In both Bauhaus and Basic Design, these cartographical approaches to form emerged after global conflict, but with different tasks in mind. While Moholy-Nagy's intentions were essentially to create a unifying, even healing, language of shared creativity in the wake of World War I, the revival of this grammatical approach by Basic Design pedagogues was to promote interdisciplinarity and adaptability in an increasingly technologized culture.[58]

For Ascott and his classmates, the work produced during the Basic Design programme settled around the two poles of form and process, in each case viewed through the lens of the increasingly systemized biology of the post-war period. An ideology of form as process was increasingly evident in both staff and student work.[59] It is therefore little surprise that Ascott and a number of fellow students developed an interest in process, interaction and change. Ascott's work from his early student years often referenced seeds or eggs because these forms unfold into possibility; they are compact symbols of fertility, growth and the generation of new life. This theme in Ascott's work reflected that at King's there was still a lot of creative engagement with issues of biomorphic form and, particularly, structures of growth and the enduring legacy of D'Arcy Wentworth Thompson's *On Growth and Form*.[60] Ascott's egg, seeds, cup and spirals represented organic form as process, in reductive visual language. A cup, for example, might be a womb, a nest, a shell or a vessel, each holding, enclosing or protecting:

> I was interested in what I call meta-forms. The idea was that one image could be generative of a cluster of meanings – bottle/womb/container/nest – for example. [...] So, take the comma shape: it's about the spiral, which is the start of everything. But the instinctive thing comes in when you play with it; so it's an unfurling in the womb, and the pause between two parts of a statement.[61]

The deep simplicity of this bio-iconography allowed for the interplay of possibilities, like routes or paths of meaning networked around the symbols. Ascott was partly inspired by the polysemy of the Hebrew language, in which a single symbol held multiple possible and related meanings according to its contextual use.

This philosophy of form as potential formed the core of Ascott's early practice, enclosing the rich and overlapping territories of mechanics, biology and process that were evident in Basic Design teaching. As a student and in the few years directly after he finished training, Ascott developed a core of ideas that were to fuel his practice throughout his career and, I argue, continue to do so. His fine art training was steeped in the fluid, metamorphic language of interwar biological abstraction, in which the intricate mechanics of growth in the living world became form. As we have seen, biomorphic abstraction held an important place within the curriculum at King's College, with regards to both Pasmore's more intuitive organicism and Hamilton's increasingly mechanical approach to biological form.[62] Over the course of his training and in the years that immediately followed, Ascott began to evolve an idea that would remain part of his ideology of practice throughout his career. This was the treatment of form as embedded in process, or in change. By the late 1950s Ascott was building a behavioural vision of the living world, derived from the implicit potential for change held within biological form. It was an idea that built upon the language of abstract form he had been taught as a student, but with a slight shift of focus Ascott had moved from form to potential, while remaining interested in biological process. This stage of his career has not received the attention it deserves, overshadowed by the enticingly unifying place of cybernetics in his practice.

While Ascott was on the cusp of the most vital period of development of his career, he was already creating links between form, process and behaviour that were steeped in the increasingly interdisciplinary debates about organismic behaviour. Ascott's cybernetic model of practice and pedagogy did not spring fully formed from his exposure to early theories, but rather from the rich culture of biology, technology and abstraction outlined in the previous chapter melding with his own background of military service. Similarly, cybernetics did not germinate whole and new from the ashes of the war. It was a product of the same debates that increasingly dominated contemporary art – the interface of systems, both living and technological.

Motion-process

There was a palpable trend amongst King's College students towards work with movement: kinetic sculpture, sculpture and painting with moving elements designed to be manipulated and works of art created through throwing, dropping or splashing.[63] In a short memoir of his student years at King's College, former student John A. Walker described creating a construction that was 'an oval shape that was hinged so it folded like a book', designed to be touched and moved, and he recalled that many students were 'interested in art that exemplified change'.[64] This statement bears an imprint of the concept that Ascott had formalized in his 1964 article 'The Construction of

Change', in which he outlined his teaching model in relation to his belief that his 'Change Paintings and kinetic constructions' were 'not only intended to discuss and project ideas, but as analogues of ideas – structures which are subject to change and human intervention in the way that ideas are'.[65] In this way, the Basic Design drive towards creating adaptability manifested itself in the unanticipated outcome of an increasingly interactive trend in the work of the students. It is also worth noting that Pasmore himself had, in the early 1950s, produced a series of paintings in the distinctive oval shape Walker remembers using.[66] The oval is inherently organic, recalling the egg and seed forms that were also important to Ascott during his studies. Edward Juler has observed that 'Modernists would see the egg shape as a symbol of the illimitability of creation and – in response to the neo-vitalist theories of Hans Driesch and Henri Bergson – the spatio-temporal extension of contemporary sculpture as a token of its burgeoning vitality'.[67] Ascott recalls that he also created oval paintings and constructions in this period, including a mural commission for the library at Newcastle.[68]

For students and teaching staff, the premise of Basic Design offered both opportunities and difficulties; it was clear that the schism between secondary and higher art education was a significant one and that it was difficult to conceive of a foundation for art as it was a rapidly evolving discipline. Ascott's early work as a student was discussed briefly by Edward Shanken, noting Ascott's self-acknowledged debt to Pasmore: '[...] Pasmore's theoretical approach to art and his agglomeration of diverse aesthetic, philosophical and scientific ideas foreshadowed the sorts of associations that Ascott would apply in synthesizing his own method.'[69] Pasmore's approach used biological form and process as a model for abstraction, forged links between art and architecture, biology and technology. While Ascott used a different model for his own pedagogy, he maintained this synthesized approach. Ascott also highlights the importance of Lawrence Gowing's passionate interest in Cezanne. Ascott wrote his dissertation about Cezanne, having drawn from Gowing's large book of facsimile prints.[70] Cezanne's work was not, in Ascott's view, about the 'representation of space', it was 'about the representation of time'.[71]

Change paintings: form, material and intention

During his three years of training at King's College, Ascott experimented with constructivism and with abstract expressionism, gradually resolving the apparent distance between the two into a kind of duality. Ascott recalled the fear of moving into different modes of practice, particularly when one style of work had been well-received. It is an interesting measure of the strange materiality of abstraction, in that constructivism and expressionism appear to occupy opposing intellectual territories: one formal, considered and planned, the other fluid, expressive and intuitive. This is a curious fallacy of materials; it is equally possible to be expressive while creating hard edges and definite forms as it is with a brush dipped in paint, although the latter is more direct.

This duality of form represents the curious notion that geometry is more intellectually considered than fluidity. The fluidity of abstract expressionism had granted it a mysticism derived from the idea that it reflected to varying extents, the artist's psyche. Construction, on the other hand, had been classified as measure of the artist's intention. This presumption apparently stemmed from the relation of straight edges and whole forms to geometry, and thus to reason. This dichotomy of construction and expression was ingrained in Basic Design teaching from the outset. In an unpublished essay, Anton Ehrenzweig offered a formal critique of Basic Design and commented on the psychological drives of order and disorder that were so intimately connected with constructive/expressive modes of abstract art. In this early essay, Ehrenzweig explored a number of Basic Design techniques and their benefit for student development, before outlining two distinct methods within them:

> I am contrasting the disruptive method aimed at loosening conscious control with another constructive kind of basic design teaching that strengthened conscious control and 'good' taste. These opposing trends have become associated with two artistic movements described by William Johnstone in his introduction; the classical Bauhaus tradition would aim at clean, aesthetic construction, while today's disruptive 'accidental' techniques appear associated with contemporary tachism or action painting.[72]

Ehrenzweig viewed exercises rooted in formal practices of abstraction as forms of behaviour – linked to specific psychological outcomes for the students. He claimed here that *Disruptive* Basic Design exercises were orchestrated by tutors to undermine processes and create surprises. *Constructive* exercises worked towards a Bauhaus ethic. This strain of behavioural pedagogy revolved around action and reaction, a trend that was manifested again in the increasingly interactive work produced by students at King's College.

It is likely that Ehrenzweig's essay was intended for a publication about the Central School, which meant it would have been necessary to present a positive point of view about Basic Design.[73] This could explain the contrast with Ehrenzweig's later opinions on the subject. *The Hidden Order of Art*, which was published posthumously in 1967, explored the rich territory of abstract art and psychology. On Basic Design, Ehrenzweig wrote:

> When Basic Design was first introduced in the days of the *Bauhaus* and the renaissance of the *Bauhaus* tradition during the nineteen-fifties it did not arrive as the desiccated analysis of empty form which it largely is today.[74]

It is clear that like most of the individuals associated with Basic Design, by the middle of the 1960s Ehrenzweig had acknowledged the limitations of the movement. However, reading this against his earlier essay, we can surmise that Ehrenzweig had recognized the strengths of the dual tactics of disruptive and constructive activity.

This duality of tactics of positive form and induced chaos is built on some base presumptions of how modes of production reflect the control – intellectual or emotional – exacted by the artist over the work of art. Given that many strains of abstract expressionist painting relied heavily upon pre-planned systems of production, this uneasy binary of construction and expressionism is an oddity. Ascott acknowledged, for example, the importance of seeing the Jackson Pollock exhibition at the Whitechapel Gallery in 1958, subsequently spending the summer experimenting with more painterly techniques than the Perspex and wood constructions which had previously occupied him. A number of the students had visited the exhibition that summer and Lawrence Alloway subsequently came to King's College to give a lecture on Pollock for the students.[75] After a summer of experimental painting, Ascott had worried about how this new gestural turn in his practice would be received upon return to King's College, recalling:

> I thought things were going to be especially tricky because my relief work had previously been favourably received by Victor, and suddenly, instead of constructions in Perspex and wood I was bringing these huge gestural paintings to the crit. I thought, what's going to come of this? But Gowing was terrific. He had recently done a tour of the States and he said, 'This is exactly what you've been doing previously. It's still constructed; it is just the other side of the coin.'[76]

This recollection brings to life the inspirational place that Lawrence Gowing held in Ascott's education. He was then Professor of Fine Art at King's College. In Ascott's view, this exchange reflected the skill that Gowing had for cutting to the core of an issue in concise and insightful terms. He highlighted that while material and method in forms of abstraction often differ, they may still be connected to the same purpose and the same outcome. In the broad spectrum of abstract form, a fluid gestural mark or a flatly painted geometric shape are different means to the same end. In this way, he resolved Ascott's gestural mark-making as another element of the same essentially formal process of creating art objects. This eased Ascott's anxieties about his shift in process, while placing gestural painting as a form of construction.

It is worth noting that Pasmore had also, at several points in his own career, attempted to reconcile construction and painting, which had always been something of an uneasy relationship for him. Pasmore was also troubled by the idea that certain aesthetic forms were more organic than others;

in his passionate exchange of letters with the American artist Charles Biederman, he took up the issue of the straight line:

> The idea that the curve belongs to nature and the straight line to man, seems to me extremely unscientific. [...] In any case, why are so-called inorganic characteristics in nature, like the straight line, legitimate for the pure painter/sculptor and not organic ones?[77]

Ascott's uncertainty hinged upon this difficulty – the associative notion of intellectual control that came with the hard line and the constructivist method as somehow in contrast to nature and to natural form.

After Ascott finished his degree he was appointed to the coveted post of studio demonstrator for a further two years. These were apprenticeships of sorts; securing this role was dependent on the approval of your former tutors and their expectations that you might go on to teach. He was well-regarded within the department – Ascott recalled that 'I was their guy [...] I got Basic Design'.[78] The fact that Pasmore took him on as a studio demonstrator was a measure of this: these posts were competitive and prized as a kind of apprenticeship to teaching.[79] It was Pasmore who secured him the post at Ealing, where it was anticipated he would teach a foundation course much in the model of Basic Design at King's College. There is a line of development between the increasingly mechanized interpretations of biological form that were produced by students and staff at King's College and Ascott's progression to cybernetic art practice and pedagogy. He was educated at the height of abstract pedagogy in British art schools, offset against the vital convergences of biology, technology and engineering of the same period.

Ascott recalls Pasmore's influence as vital to his own development and Hamilton's influence of equal, if less direct, value.[80] It was during this period of employment that Ascott made his first *Change Paintings* in 1959–61, while he was a studio demonstrator at King's College (Figure 1.9).

The paintings consisted of abstract marks on movable Plexiglas squares which could be rearranged into endless configurations, overlapped or re-ordered. Ascott offset the precision of his construction of sliding panels against the rawness of gestural marks painted in dragging brushstrokes of thick, black paint. The *Change Paintings* played upon the same tension that Gowing had alleviated for Ascott – the intuitive and the formal. The gestural marks themselves bear an imprint of Pasmore's methods; in classes he would often demonstrate free mark-making for his students, lines of marks appearing like ambiguous hieroglyphs. His student audience witnessed the relationship between mark and process, a gestural dance that resulted in analogous forms on canvas. Ascott's marks have that free quality, the dripping black paint in contrast with the clean, and very contemporary, surface of the Plexiglas. He described these marks as 'seeds' from which the composition could grow through the movement of the Plexiglas sheets.[81] Ascott viewed these paintings as an essential phase in his development as they can be manipulated and therefore have the analogue

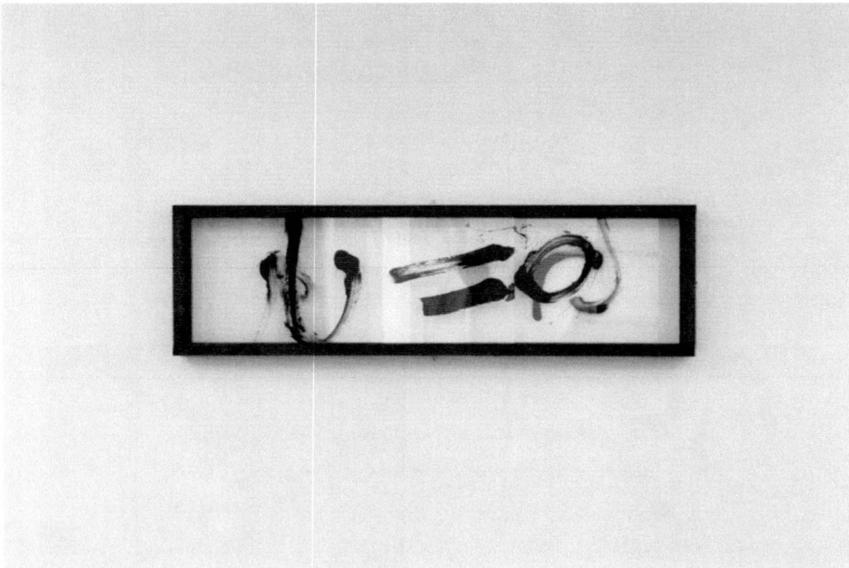

Figure 1.9 Ascott (1959) *Change Painting*. Wood, Plexiglas and paint. 66 × 21 in

qualities that dominated both his work and his pedagogy throughout the 1960s.[82] Their interactive quality turns the paintings into games, experiences, even performances. His classmate John Walker recalled Ascott's interest in the spectator, writing that the *Change Paintings* '[...] included interchangeable elements that encouraged spectator participation and literally exemplified change'.[83]

The materials used in these works reflected the boom in new fabrication materials that had been engineered for military applications and were, after the war, put to other uses. In this way the common use of branded acrylic products such as Perspex and Plexiglas in the visual art of the 1950s and 1960s was in part provoked by an increase in production and application during World War II, when they were used for aircraft windshields and submarine periscopes, amongst other military applications.[84] Plexiglas was proven to cause less injury when it shattered than glass. Acrylic products segued from the military to the domestic after the war, as did hundreds of other products that benefited indirectly from research into their more sophisticated application during the war. These base materials became the architectural fabric of modern life. They were also popular with artists, including both Ascott and his then teacher, Victor Pasmore. Pasmore's *Linear Motif in Black and White* of 1960–1 (Figure 1.10) is contemporaneous to Ascott's *Change Paintings*, and uses two materials that had also been used to build aircraft during World War II – Formica and plywood. It is one of a small number of similar constructions that Pasmore created, including the similar work in the Tate collection.

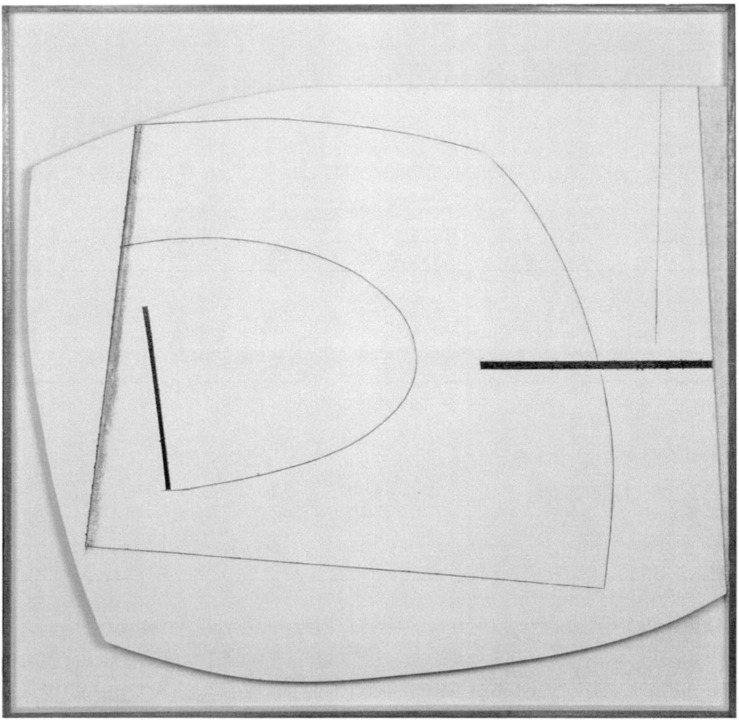

Figure 1.10 Victor Pasmore (1960–1) *Linear Motif in Black and White*. Oil paint on Formica and plywood. 1219 × 1219 mm. Photograph courtesy of the estate of Victor Pasmore

Pasmore's use of these modern materials – Plexiglas, Perspex, plywood and Formica – reflected a trend on both sides of the Atlantic, as materials engineered for modern warfare were absorbed into design and industry. These were the materials of the modern home: clean, pure and untrammelled by history. This is despite their recent development in answer to the needs of the military. Pasmore recalled having a '[...] naïve notion about using modern synthetic materials as a concession to modern technology. But modern technology is electronics not Perspex!'[85] His 'naïve' notion of the modernity of these materials was, however, correct insofar as modern materials were the product of modern electronics, engineering and chemistry. However, Pasmore had a purist approach to material, and the associations of new materials – military, modern and otherwise – were not so important to him as their aesthetics:

> For me relief construction was an experimental development intended to extend the reality of flat painting, and its method was empirical. The use of an extended fin and a transparent plane was intended to create

optical ambiguity and direct the eye beyond the boundary of the rectilinear format.[86]

This certainly differed from his students, including Ascott, who dwelt upon the philosophical implications of extending from the picture plane into three dimensions, and the effect this disruption might have on the viewer.

Many of these materials were used for forms of construction-based abstraction, and because of this, post-war abstraction had an aesthetic that was both modern but also tied, intimately, to the military–industrial complex that held such a vital and problematic role in the arts after the war. Indeed, these materials are anything but bland and modern; they reflect the strange and troublesome relationship of conflict and progress, as well as a strain of techno-positivism that was manifest in the visual arts.[87] In the *Change Painting* that featured in the catalogue of Ascott's 1963 exhibition at the Molton Gallery in London, the surface patterns included electrical circuits connected by a thick, coiling line (Figure 1.11).

This application of electronic diagram as pattern demonstrates Ascott's growing interest in systems thinking in the early 1960s, the vital years that would see him take interactivity in new directions using analogues, play and eventually, computers.

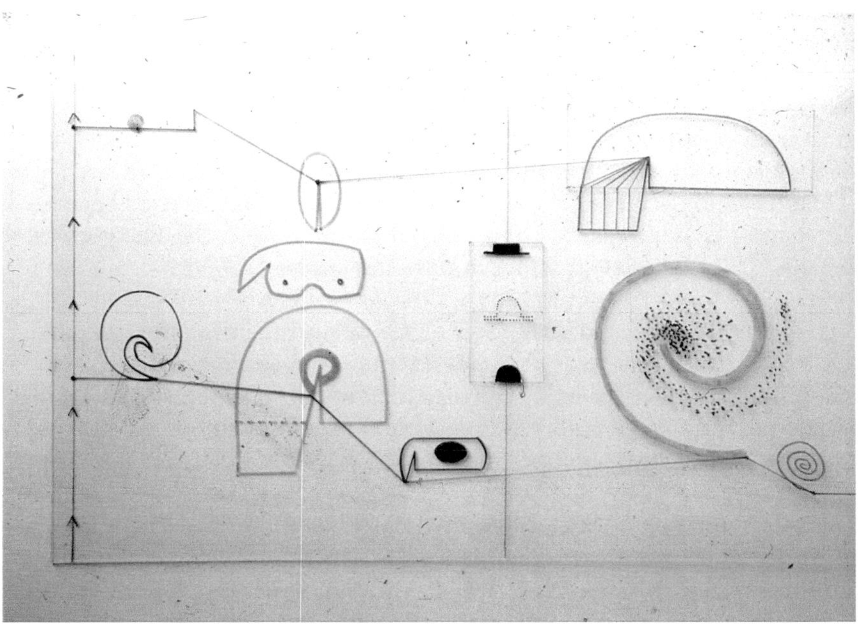

Figure 1.11 Roy Ascott (1963) *Drawing on Perspex (for Analogue)*. Photograph courtesy of Roy Ascott

The modern materials that Ascott and his contemporaries used also spoke of recovery – for the duration of the war, the possible commercial applications of new materials had been supressed by the prioritized manufacture of military supplies of all kinds.[88] For example, Robert Friedel has described how the plastic melamine formaldehyde had been used to manufacture helmet liners for US soldiers and after the end of the war, the company American Cyanide collaborated with the designer Russel Wright to create the first widely available melamine tableware.[89] In this way, the liberation of these materials arose from the end of the conflict. Given that many of these new materials had only ever been used in applications connected to warfare, their post-war development in industry and manufacture drove a spate of innovation in design and building.

With this in mind, we might reassess the application of new materials in British abstraction as part of a new aesthetic of the modern, particularly in light of their dominating presence in architecture, interior design and home ware. Sam Gathercole has noted that Pasmore had been:

> [...] keen to work in social spaces, accepting several public commissions. In work for Newcastle University (1956) and Fairlawn Primary School in Lewisham (1958), Pasmore aimed at an 'objective' synthesis in which 'unitary principals governing the 'structural processes' were common to both art and architecture.[90]

For Pasmore, new materials were perfect for the politics of social regeneration in post-war architecture, embodying modernity and democratizing social space. It was during the development of these projects that Pasmore had been in correspondence with Biederman, debating the definition of 'structural processes.'[91] The words 'structural processes' are essentially a summation of the debates that had arisen around the *Growth and Form* exhibition and the *Aspects of Form* symposium. As Thistlewood has noted, Alfred North Whiteread's philosophy of form was at the heart of these debates.[92] He writes of structure in time and space:

> [...] Whiteread held that the essence of an organism was the changing pattern of its total structure extending in space and time. Thus to Whiteread reality was a structure of evolving structures; and the particular instance of reality was structural processes embodied in the subdivision called the event.[93]

This has sympathy with Ascott's own early identification with Cezanne's paintings as depictions of time rather than of space. Pasmore was, during the time of Ascott's studies at King's College, working through concepts of synthesis, process, construction and architecture, extending his practice into more directly social projects such as architectures for education. This overlap adds another dimension to the use of materials in art that were fast replacing traditional materials in architecture too. Ascott's *Change Paintings*, then, were part of a process of absorption in which the application of newly engineered materials contributed to a future-facing aesthetic for abstraction, particularly in the context of positivist constructivism. Ascott had also experienced a unique exposure to the materials in their military applications, given his service in Fighter Control, where the engineering, the communication networks and the very modern materials in themselves left an indelible imprint upon his psyche.

Shanken has discussed the *Change Paintings* in relation to the creative interest in mechanical biology inspired by D'Arcy Wentworth Thompson which was prevalent in the Basic Design curriculum, describing how for Ascott each panel 'was a painterly gesture that the artist conceived of as a "seed" or "ultimate shape"'.[94] Some of Ascott's later *Change Paintings* were more figurative and employed a sharper, more graphic style, in a palette of red and black (Plate I).

The female torso painted in black on the left echoes in the split wedge painted in red on the right, the genitalia depicted as an opening to the enclosed, fertile whole of the lower torso. The style recalls Joan Miró's reductive depictions of the fertile female body (Plate II), although Ascott's bodies are less aggressive towards female reproduction than Miró's diagram of motherhood: a body reduced to its reproductive elements, feeding parasitical and insect-like babies.

In the centre of this *Change Painting*, a black zigzag line and three black dots are divided by a series of red horizontal lines, bringing to mind fertilization. Ascott's metaforms were intended as a language of growth and reproduction, organic symbols of possibility. In this way, Ascott aligned the mechanics of biological growth with the mechanics of interaction; it was through movement resulting from physical manipulation that the fertile possibilities for change would play out.

Shanken also noted that 'Ascott's *Change Paintings* and later kinetic works added a durational, interactive element that further liberated form and color by allowing it to move and change over time'.[95] The suggestion that form and colour could be 'liberated' through interaction is compelling, connecting as it does to an idea that painting and sculpture were trapped in the confines of their media, frozen upon completion. Set against the larger debates on the boundaries of the art object that took place during the 1960s, Ascott's interests hold parallels with the subject–object debates of Minimalism. Robert Morris wrote that: 'the major aesthetic terms are not in but dependent upon this autonomous object and exist as unfixed variables that find their specific definition in the particular space and light and physical viewpoint of the spectator.'[96] The instability of the art object, how it changes according to environmental factors, was one of the fundamental questions that connected disparate art practices in the 1960s. Ascott's gradual development from process to interactivity hinged around the same questions of the boundaries of an object and its place within wider systems of communication. At the beginning of the decade, Ascott was also working progressively towards a more systemic view of the art object, writing in 1960 that:

> Interchangeable elements, each with an individual identity, may, by the physical participation of the spectator, be brought into a series of relationships, each one adding up to a whole which is more distinctly relates to the manipulator of the parts, than if it were static and at a distance. The act of changing becomes a vital part of the total aesthetic experience of the participant.[97]

These words were published in a pamphlet for Ascott's 1960 exhibition, *Change Paintings and Reliefs*, at St. John's College in York. In this pamphlet, Ascott's gave a systematized account of the conditions of art viewing in relation to change, emphasizing that the object was not an endpoint but an open-ended process. Even before Ascott had encountered cybernetic theory, he was certainly beginning to articulate a systems view of art practice; indeed, the prolificacy of systems references in art theory and criticism of the 1950s and 1960s indicated the way in which systems principles had already been absorbed into the visual arts.

Ascott's early work on the territory of abstraction, process and interactivity brought technology and biology together seamlessly, demonstrating his early interest in the universality of systems behaviour. It was a feature of his practice as he brought together the biomorphic abstraction that had been a cornerstone of his training in art with new theories of communication derived from cybernetics, behaviourism and engineering that had been the cornerstone of his military training. The gap between Ascott's interest in the organic and the technological was narrowing across disciplines as a more systematized age began that replaced the reductionism that had dominated the sciences. Life, in the new understanding, was process; the organism, like the computer, is process not product. Haraway writes: '[...] one is not born an organism. Organisms are made; they are constructs of a world-changing kind.'[98] Transmuting the mechanical conception of organismic growth and change that had been developed and debated by biologists throughout the twentieth century into metaphor, Haraway positioned biological change as a construct. Certainly, the idea of the organism as process and in symbiosis with their environment was at the core of early British cybernetics. When Ashby wrote in *Design for a Brain* that 'The organism affects the environment, and the environment affects the organism: such a system is said to have 'feedback', he established that living systems, like his homeostatic systems, adapted in order to remain stable.[99] They were, as Haraway deems it, 'constructed'.

In his exchange of letters with Biederman, Pasmore had also argued against the idea that the artist's response to nature was imitative. His somewhat romantic view of the artist's place in nature has an underlying message of interconnectivity, and of life – and art – as process:

> [...] the artist is a creative as well as an imitative being. He is not compelled to copy nature because he himself is a process of nature containing the very same elements which he sets out to copy. In other words he can produce the likeness of nature out of himself. His mind is conditioned as much by its own structural process as by the processes imposed on it from the outside; hence its power of projection.[100]

This statement reflects one of Pasmore's prevalent critical issues with Biederman, which was that Pasmore believed in the equal importance of intuitive, personal modes of work to rational, constructive practice. For Pasmore, intuition was closer to nature, an idea that he believed was present in the practices of Bauhaus Pedagogues Arp and Klee.[101] A similar perspective informed Herbert Read's writing upon abstraction and the organic; he commented that abstract artists dealt with '[...] certain proportions and rhythms which govern growth, including the growth of the human body. Attuned to these rhythms and proportions, the abstract artist can create microcosms which reflect the macrocosm [...].'[102] Thistlewood summarized this dominant biological analogy for process in art in relation to Read, Whiteread and Whyte:

> Organic development in a work of art is at least analogous to, and probably identical with, organic development in nature; in an organic-artistic scheme the essence of art is in processes rather than its products; and such artistic 'events' as *are* thrown up are significant merely in that they reflect past, present and future aspects of the dominant process.[103]

In Read's microcosmic metaphor, the work of art was as a single organism which could reveal the vast ecological system of which it was part. As Thistlewood has noted, in the introduction to *Aspects of Form,* L.L. Whyte had articulated the increasingly prevalent comparisons artists were expressing between abstraction and mechanical biology.[104] Whyte suggests: 'May not biological organization and mental activity be names for processes of closely similar, or even identical character, operating in different situations?'[105] Thistlewood traces Whyte's ideas in the context of the interest in the processes and systems of the human brain which was apparent in discussions at the ICA in the 1950s.[106] In the integrative artist-in-nature model Pasmore articulated at several points in the early 1950s, there is the same sense of flow – form as process, and the artist as part of this process, both psychologically and physiologically. Herbert Read had written in a similar vein in 1951, in *Art and the Evolution of Man*, when he stressed: 'The point I am trying to make, the whole point of my hypothesis, is that the work of art is not an analogy—it is the essential act of transformation;

not merely the pattern of mental evolution, but the vital process itself.'[107] He had, in 1947, described art making as an essentially biological imperative, described as a product of fertility:

> Art is human, not divine: profane, not sacred. It does not descend in Pentecostal flames: it arises, like a green sap; like a seminal fluid, it issues from the body, and from the body in an unusual state of excitement. This is true whether we are literal, and think of the body of the individual artist; or metaphorical, and think of the body of society.[108]

These interplays between process in biology, psychology and creative process in turn problematized the foundational principle that a work of art is 'finished'. The questions that Haraway asked, in 1976, of biological form, are useful here:

> Form is more than shape, more than static position of components in a whole. For biology the problem of form implies a study of genesis. How have the forms of the organic world developed? How are shapes maintained in the continual flux of metabolism? How are the boundaries of the organized events we call organisms established and maintained?[109]

The question of maintenance of form crosses between the strains of abstraction discussed here and the biological processes that inspired them. While the development of a work of art and its eventual form can be traced, it is harder to define how this form is maintained, or fixed in place. 'Continual flux', process, flow – boundaries had become elastic, concepts rather than actualities.[110]

It is becoming increasingly evident that the major shifts in biological–mechanical communication during and after the war had a larger part to play in the early phases of conceptual art than has hitherto been acknowledged. Further, introducing systems thinking into the theorizing around object and subject tore through the etymological dualism of the terms. The Latin root *objectus* means to throw before or against, while *subjectus* is to be brought or thrown under, or influenced. The classical understanding of an object is thus that of resistance – something that we can sense or touch that will resist being absorbed into our own selves. Once the arbitrary and philosophically problematic notion of the separateness of the art object as a quality to be measured in an act of distinction is abandoned, then art itself dissolves into its environment, and into time. As Pamela Lee so eloquently argued in *Chronophobia: On Time in the Art of the 1960s*, disruptions, contractions, leaps and shifts in time were an intrinsic element of the profound and complex changes produced by a technologising culture:

> I treat the obsession with time in 1960s art in tandem with two indissociable shifts in the culture following World War II: the alleged waning

> of the 'Machine Age' on the one hand, and the concomitant advent of computer technologies, on the other. I suggest that the rise of the information age and its emphasis on speed and accelerated models of communication serve as the cultural index against which many artists and critics gestured.[111]

Communication technologies certainly warp our conceptions of time; messages in the form of disembodied information can be transmitted instantly, transformed into other kinds of data, fed from human to machine and back again. Time is part of the fabric of a seismic change in communication, and I contest that this could be deemed a cultural index; culture itself was in flux, enfolding artists rather than forming a backdrop to practice. Technology is not distinct from nature, and as such Pasmore's claim that art 'imitates nature in its manner of operation' has new relevance.[112] As Ascott himself emphasizes, technology is not the source. It is created from human research and ambition and like and other inventions in our world, it uses the laws of nature in order to further the reach of the human hand and mind. 'Technology', Ascott tells us, 'is the product of desire'.[113] Ascott emphasizes that technology should not be seen as the focus of his career, despite his consistent engagement with the most modern of communication technologies since his student years. It is communication itself, its flux and flow, that has endured as a passion – that and its reflection of the infinitely mysterious field of human consciousness.

Notes

1 As noted in the introduction to this book, several Ratio Club members came from a psychiatric background, including W. Ross Ashby and William Grey Walter.

2 For a full account of field theory in biology, see: Donna Haraway (1976) *Crystals, Fabrics and Fields: Metaphors that Shape Embryos*. Berkeley: North Atlantic Books.

3 The Central School of Arts and Crafts later merged first with the London Institute and then with Central St Martins in 1989 to form Central St Martins College of Art and Design.

4 King's College was based in Newcastle, although it was part of the University of Durham. King's College became the University of Newcastle upon Tyne in 1963.

5 Ascott has acknowledged Pasmore's influence numerous times. See, for example: Shanken 'From Cybernetics to Telematics: The Art, Pedagogy and Theory of Roy Ascott' in: Ascott & Shanken (2003) *Telematic Embrace: Visionary Theories of Art, Technology and Consciousness*. Berkeley: University of California Press. 1–94, 9; and Roy Ascott in Meeka Walsh 'Analogue: Incidents in the Life of Roy Ascott'. *Border Crossings*, Issue 127, August 2013.

6 In 1970 and until Pasmore's death, several negative or ambivalent reviews of his exhibitions were published. See, for example: Jane Rye (2010) 'Review: Victor Pasmore: Writings and Interviews'. *The Burlington Magazine*. Vol. 152. No. 1290, 623; Frances Spalding (1980) 'Liverpool: Victor Pasmore at the Walker Gallery'. *The Burlington Magazine* Vol. 122, No. 925, 278–81, 278;

and Andrew Causey (1999) 'Victor Pasmore. Liverpool and London'. *The Burlington Magazine* Vol. 141. No. 1158, 565–7.

7 For the most detailed account, see Richard Yeomans (1988) 'Basic Design and Richard Hamilton's Teaching'. *The Journal of Art and Design Education.* Vol. 7. No. 2, 155–6.

8 Hamilton, R. (1949) *Growth and Form* Proposal for Herbert Read. Tate Gallery ICA archive. TGA 955.1.12.26. Also reproduced in Jurgen Jacob. (1986) *Die Entwicklung der Pop Art in England.* New York: Peter Lang.

9 Richard Hamilton (1950) *Growth and Form Exhibition, First Draft Schedule*, ICA Papers, Tate Gallery Archive 955.1.12.26, also reprinted in Jacob. J (*ibid.*) He dates this to 1949, slip page at Tate marked Dec 1950.

10 Sigfried Giedion [1948] (1975) *Mechanization Takes Command: A Contribution to Anonymous History*. Oxford: Oxford University Press.

11 Richard Hamilton (1949) *Draft Proposal for Growth and Form.* ICA Archives. p. 1. Also cited by Anne Massey (1995) 44.

12 Isabel Moffatt (2002) *The Independent Group's encounters with logical positivism and searches for unity in the 1951 Growth and Form Exhibition.* Doctoral Thesis. MIT. Available online at http://dspace.mit.edu/handle/1721.1/17555?show=full, accessed November 2011.

13 Hamilton (1982) 10.

14 D'Arcy Wentworth Thompson (1917) *On Growth and Form*. Cambridge: Cambridge University Press, 6.

15 Charles Baudelaire (1863) 'The Painter of Modern Life'. In: Baudelaire (1964) *The Painter of Modern Life and other Essays.* London: Phaidon. 13.

16 See: Edward Juler (2015) *Grown but Not Made: British Modern Sculpture and the New Biology*. Manchester: Manchester University Press, and (2013) 'A Bridge between Science and Art? The Artistic Reception of *On Growth and Form* in Interwar Britain, c. 1930–42'. *Interdisciplinary Science Reviews*, Vol. 38. No. 1, 35–48.

17 Thompson (1917) 6.

18 *Ibid.*, 17.

19 Giedion (1948) 718.

20 Giedion (1948) 41.

21 S.J. Gould (1971) 'D'Arcy Thompson and the Science of Form'. *New Literary History*, Vol. 2., No. 2. Form and its Alternatives. Winter 1971, 229–58. 10.

22 Thompson (1917) 11–12.

23 Ludwig Von Bertalanffy (1969) *General System Theory: Foundations, Development, Applications*. New York: George Braziller, 48.

24 Thompson (1917) 9.

25 Thompson (1917) 232.

26 Thompson (1917) 233–4.

27 Herbert Read (1951) 'Preface' in: Whyte, L.L., *Aspects of Form*, xxi.

28 *Ibid*. 2.

29 *Ibid*. 2.

30 Kathleen Lonsdale (1953) 'Aspects of Form' (book review) *Acto Crystallographica*. 224.

31 Whyte. *Op Cit*. Joseph Needham (1951) 79.

32 Elena Crippa and Beth Williamson (2013) (eds) *Basic Design*. London: Tate Gallery, 8.

33 Victor Pasmore (2001) In: Nicholas Watkins. 'An Interview with Victor Pasmore'. *Burlington Magazine*, Vol. 143. No. 1178, 284–9, 287.

34 *Ibid*. Pasmore (2001) 287–8.

35 Richard Yeomans, *The Pedagogy of Victor Pasmore and Richard Hamilton*. Typescript of paper delivered at the Henry Moore Foundation on

Nov 4 2009. Available online at http://www.henry-moore.org/docs/yeomans_basic_design_0.pdf. 9, accessed 01/11/11.

36 While Yeomans compares Pasmore's approach to Kandinsky's notion of the 'art laboratory' at the Bauhaus, it developed more from Pasmore's own relatively recent conversion to abstraction.

37 As a student in in late 1940s, Hamilton read Siegfried Giedion *Mechanization Takes Command* (1948), and then completed his *Reaper Series* of prints.

38 Grieve (1982) 546.

39 Richard Hamilton (1982) *Collected Words 1953–1982*. London: Thames and Hudson. 12.

40 See Dawn Leach (1993) *Richard Hamilton: The Beginnings of His Art*. Frankfurt: Peter Lang; Anne Massey (1995) *The Independent Group: Modernism and Mass Culture in Britain, 1945–1959*. Manchester and New York: Manchester University Press; Isabelle Moffatt (2002) *The Independent Group's Encounters with Logical Positivism and Searches for Unity in the 1951 Growth and Form Exhibition*. PhD Diss. MIT; David Robbins (1990) *The Independent Group: Postwar Britain and the Aesthetics of Plenty*. Cambridge: MIT Press; and Graham Whitham (1986) *The Independent Group at the Institute of Contemporary Arts: Its Origins, Development, and Influences 1951–1961*. PhD Diss., University of Kent at Canterbury.

41 Oliver Botar & Isabel Wünsche (2011) *Biocentrism and Modernism*. Ashgate. Aldershot, 1.

42 C.P. Snow (1959) *The Two Cultures and the Scientific Revolution*. London: Cambridge University Press.

43 Sigfried Giedion [1948] (1975) *Mechanization Takes Command: A Contribution to Anonymous History*. Oxford: Oxford University Press.

44 Sigfried Giedion (1941) *Space, Time and Architecture: The Growth of a New Tradition*. Cambridge Harvard University Press.

45 Albert Einstein cited by Arthur P. Molella (2002) 'Science Moderne: Sigfried Giedion's *Space, Time and Architecture* and *Mechanization Takes Command*'. *Technology and Culture*, Vol. 43, No. 2, 374–89, 377.

46 *Ibid*. p. 378.

47 *Op. Cit*. (1982) 12.

48 *Op. Cit*. (1948) 84.

49 *Op Cit*. (1982) 12.

50 Hamilton (1982) 12.

51 This has been the subject of a recent exhibition and accompanying publication: see Carson Chan (ed.) (2017) *Richard Hamilton & Sigfried Giedion: Reaper*. Zurich: JRP Ringier.

52 See Maurice de Sausmarez *et al*. (1961) 'A Visual Grammar of Form', *Motif*, Winter, 3–29 and de Sausmarez *et al*. (1962) 'A Visual Grammar of Form(2)' *Motif*, Summer, 47–67.

53 Victor Pasmore e*t al*. (1959) *The Developing Process: Work in Progress Towards a New Foundation in Art Teaching as Developed at the Department of Fine Art, Kings College Durham University, Newcastle upon Tyne, and at Leeds College of Art*. Published on the occasion of an exhibition at the ICA London. Durham: ICA & King's College, University of Durham.

54 Wassily Kandinsky (1926) *Point and Line to Plane*. Bauhaus.

55 Ellen Lupton (2008). In: Ellen Lupton & Jennifer Cole Phillips, *Graphic Design: The New Basics*. New York: Princeton Architectural Press. 9.

56 By the 1950s the Bauhaus influence in the capital was well-developed, particularly in the art schools.

57 Erik Forrest (1983) 'Harry Thubron: His Contribution to Foundation Studies in Art and Design Education'. PhD Diss., The Ohio State University. 52–3.

58 For an account of Moholy-Nagy's utopian design ideals at the New Bauhaus, see Victor Margolin (1997) *The Struggle for Utopia: Rodchenko, Lissitzky, Moholy-Nagy, 1917–1946*. Chicago: University of Chicago Press.
59 Richard Yeomans (2009) *The Pedagogy of Victor Pasmore and Richard Hamilton*. Typescript of paper delivered at the Henry Moore Foundation on Nov 4 2009. Available online at http://www.henry-moore.org/docs/yeomans_basic_design_0.pdf. 9, accessed 01/11/13.
60 D'Arcy Wentworth Thompson (1917) *On Growth and Form*. Cambridge: Cambridge University Press.
61 Roy Ascott. 'Analogue: Incidents in the Life of Roy Ascott' *Border Crossings*. Issue 127., August 2013. Available online at http://bordercrossingsmag.com/article/analogue-incidents-in-the-life-of-roy-ascott, accessed September 2015.
62 Richard Yeomans 'Basic Design and the Pedagogy of Victor Pasmore and Richard Hamilton'. In Mervyn Romans (ed.) (2005) *Histories of Art and Design Education*. Bristol: Intellect Books.
63 These material approaches are fully detailed by Ehrenzweig (1965), Pasmore *et al.* (1959) and de Sausmarez (1965). As well as a grammar of form, the Basic Design approach had an ingrained grammar or process and gesture.
64 John A. Walker (2003) *Learning to Paint: A British Art Student and Art School 1956–61*. London: Institute of Artology. 39.
65 Roy Ascott (1964) 'The Construction of Change'. *Cambridge Opinion*, 41 *(Modern Art in Britain)* 37–42, 37.
66 Alan Bowness. 'The Paintings and Constructions of Victor Pasmore'. *The Burlington Magazine*, Vol. 102, No. 686, Modern Painting (May, 1960) 198–205, 205.
67 Edward Juler (2015) *Grown but not Made: British Modernist Sculpture and the New Biology*. Manchester: Manchester University Press, 52.
68 Roy Ascott & Kate Sloan *In Conversation*. Henry Moore Institute, Leeds. 8 March 2017.
69 Shanken. *Op. Cit.* (2003) 9.
70 Kate Sloan *Interview with Roy Ascott* (2011).
71 Roy Ascott & Kate Sloan *In Conversation*. Henry Moore Institute, Leeds. 8 March 2017.
72 A. Ehrenzweig. *Psychological Factors in Teaching Basic Design*. Unpublished draft typescript. William Johnstone Archive, National Library of Scotland Dep. 332/6.
73 Anton Ehrenzweig did write the essay for a short publication produced at the Central School of Art in 1959 entitled *William Johnstone: Artist and Art Educator*. London: Central School of Art. The essay was also published as an article of the same title in *The Studio* in 1959.
74 A. Ehrenzweig (1967) *The Hidden Order of Art: A Study in the Psychology of Artistic Imagination*. London: Weidenfeld and Nicolson, 57.
75 John A. Walker (2003) 30.
76 Roy Ascott (2013).
77 Victor Pasmore. In: Alastair Grieve 'Constructionists I: Biederman and Victor Pasmore'. *The Burlington Magazine* Vol. 124, No. 954, Special Issue Devoted to Twentieth-Century Art (Sep., 1982), 540–51, 548.
78 *Ibid.* (2012).
79 Shanken (2003) described this as a 'grooming process', with the chosen students expected to take forward their master's pedagogies into new posts which were often secured for them.
80 Kate Sloan *Interview with Roy Ascott* (10 September 2012).
81 *Ibid.* (2012).
82 *Ibid.*

83 John A, Walker (2003) 52–3.
84 Alfred D. Chandler Jr. (2009) *Shaping the Industrial Century: The Remarkable Story of the Evolution of Modern Chemical and Pharmaceutical Industries*. Cambridge: Harvard University Press, 92.
85 Leif Sjöberg 'On Victor Pasmore's Use of Transparency' *The Structuralist*, Jan 1 1987, Vol. 0 (27), 88–90, 90.
86 *Ibid*., 90.
87 While the military-industrial complex has most often been discussed in relation to art and technology in the USA, we might extend the scope of this concept to enclose engineered materials, as well as the strain of positivism that was manifest in the work of some UK artists and scientists, including Hamilton, Pasmore, Ascott, Pask and Ashby.
88 Donald Albrecht (ed.) (1995) *World War II and the American Dream: How Wartime Building Changed a Nation*. Cambridge: MIT Press.
89 *Ibid*. Robert Friedel (1995) 68.
90 Sam Gathercole 'Art and Construction in Britain in the1950s'. *Art History*, Vol. 29, Issue 5. Nov 2006, 887–925, 919.
91 Alistair Grieve (1982) 546.
92 David Thistlewood. 'Organic Art and the Popularization of a Scientific Philosophy'. *British Journal of Aesthetics*, 22:4 (1982: Autumn), 331–21, 318.
93 *Ibid*. (1982) 318.
94 Shanken (2003) 29.
95 *Ibid*. (2003) 30.
96 Robert Morris 'Notes on Sculpture, Part II' *Artforum* vol. 5 no. 2, 20–23, 23.
97 Roy Ascott, statement from the exhibition brochure 'Change-Paintings and Reliefs', St. John's College, York, 1960.
98 Donna Haraway (1989) 207.
99 W. Ross Ashby (1960). *Design for a Brain*. London: Chapman and Hall. 37.
100 *Ibid*., 546.
101 Pasmore attended a symposium at the ICA about Klee's *Pedagogical Sketchbook* in 1953, where he discussed 'nature in art's processes' in relation to Arp. See Grieve (1982) 548.
102 See David Thistlewood (1998) 'Herbert Read's Organic Aesthetic 1918–1950'. In: *Herbert Read Reassessed*. Liverpool: Liverpool University Press, 215–32, 222.
103 David Thistlewood (1982) 'Organic Art and the Popularization of Scientific Philosophy'. *British Journal of Aesthetics*, Vol. 22. No. 4, 311–21, 317.
104 *Ibid*., (1982) 318.
105 *Ibid*., 318.
106 *Ibid*., 318–19.
107 Herbert Read (1951) *Art and the Evolution of Man*. London: Freedom Press, 31.
108 Herbert Read (1947) 'The Fate of Modern Painting'. *Horizon*, November, 242–54, 253.
109 Donna Haraway (1976) *Crystals, Fabrics, and Fields: Metaphors that Shape Embryos*. Berkeley: North Atlantic Books, 39.
110 *Ibid*., 39.
111 Pamela M. Lee. (2004) *Chronophobia: On Time in the Art of the 1960s*. Cambridge: MIT Press.
112 Victor Pasmore quoted in: Grieve (1982) 542.
113 Kate Sloan *Interview with Roy Ascott* (November 2015).

2 Analogue

Interactivity | performativity | cybernetics

Analogues and World War II technology

There was a brief spell in the history of technology when the analogue and the digital coexisted and interacted, spanning World War II and the decade that followed. During this interlude, Roy Ascott experienced first-hand the subtle interplay between the two at a time when the technological future was as yet undecided, and when the analogue might yet have dominated the future. One form of technology was abstracted and disembodied, the other physical and, by nature, precisely representational. For Ascott, the analogue was an intimate and perfect machine, its form shaped entirely by its function. When, in the early 1960s, Ascott began to create what he termed '*Analogues*' or '*Analogue Structures*', he had distilled the poetry of the analogue machine and applied it to sculptural form. His conception of the analogue work of art operated on two levels. He recognized that the work of art was by nature analogic – it was an analogue of the decisions of its creator. Beyond this though, as he moved into the creation of increasingly interactive work, he also explicitly described his work as analogical to the 'decisions of the viewer'.[1] Ascott's *Analogues* were exhibited at the Molton Gallery in 1963, then at the Centre D'Art Cybernétique, Paris and at Queen's University, Belfast in 1964.[2]

The term 'analogue' has its roots in the Greek *analogos*, meaning proportionate. In contemporary times, 'analogue' is most often applied as a counterpoint for digital, as with clocks – the hand that points at a number versus a digital display. In the analogue clock, the mechanism of the clock represents in a concrete way the passing of time, the hands marking out each second in an analogous movement. In this way, analogues embody meaning in a precise and entire way, with compelling and poetic resonances with the work of art. Ascott's analogues were rooted in what was the greatest technological change of the century – the communication of information from, and between, machines. Analogues, in their many wartime and post-war applications, physically calculated information and, often, physically communicated information.

The many applications of analogue technologies in the military before and during Ascott's service ranged from radar, tide-predictors, missiles, bombsights and differential analyzers (an analogue computer that could solve differential equations). Analogue computers were used at sea, such as the Ford Fire Control System which was designed in the US and purchased by the UK for a number of Royal Navy ships. These computers used sensors, thermometers and other reading equipment to measure factors including the current temperature, speed, distance and wind conditions, which assisted with the optimum deployment of weapons. The information was proportionate – analogous – to the situation, and this pure relationship had its own beauty. Analogue computers were mechanical systems; later in the war and in the post-war years these mechanical systems were replaced with electronic systems, and sensors provided the readings to the computer. For example, the British RAF Mark IX bombsight (Figure 2.1), used by British and Canadian planes during the war, had wires that had to be manually set to reflect speed, altitude and bombload – if any of these factors changed then it had to be manually adjusted. Over the course of the war, self-adjusting bombsights were created by connecting sensors that took these readings. In this way, analogues were ever-more complex as they functioned within systems of mechanical communication.

Analogue machines played a vital part in how the war was fought and won – it was also during the war that the first innovations were made in digital computing, from digital/analogue hybrids to the first electronic fully programmable computers. In this sense, we can recognize that the analogue was an increasingly complicated mechanism, made so by the complexity of systems behaviours that were manifest *between* analogues. The field of cybernetics emerged during these war years, during which scientists, engineers and psychiatrists were all deployed to fight what Churchill later described as a '[...] secret war whose battles were lost or won unknown to the public; and only with difficulty is it comprehended, even now, by those outside the small high scientific circles concerned. No such warfare had ever been waged by mortal men.'[3]

In the United States, Norbert Wiener named the field of cybernetics after working with servomechanisms in the war – these are machines that use error-sensing negative feedback to correct a system's function and they were

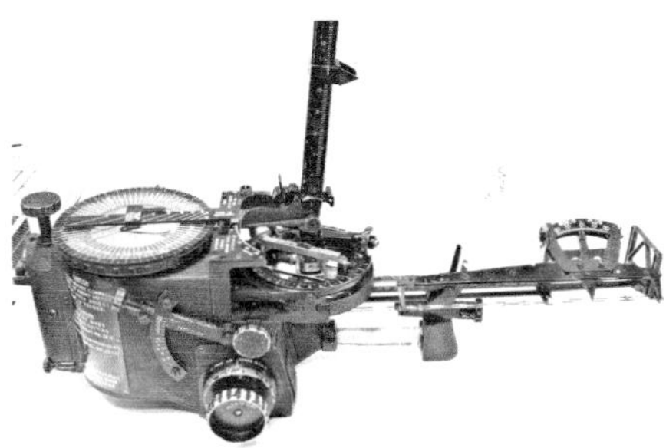

Figure 2.1 British RAF Mark IX bombsight. Photograph: Glenns Museum

employed in military fire control and marine navigation equipment. The word 'servomechanism' is derived from the French term '*Le Servomoteur*' or 'Slavemotor'. In response to working with servomechanisms, Wiener coined the term cybernetics, derived from the Greek (*kybernētēs*), meaning 'steersman, governor, pilot, or rudder'. From slave to steersman, the core issue for cybernetics was that of controlling or giving direction to complex systems, both human and technological. The etymology of the two terms highlights the anxiety of the rapid technological development that took place during and after World War II and the desire to understand and harness the increasing power and sophistication of mechanical systems. Immediately following the war, the analogue operated around a curious duality: self-contained and dedicated to a single task, yet increasingly interconnected, the information it produced communicated forwards within a mechanical system. An analogue was by nature limited to the task it performed, remaining a true measure of a pre-determined quantity.

British cybernetics, psychiatry and the machine

In the decade that followed World War II, scientists, engineers and psychiatrists converged upon the field of cybernetics in a theoretical and practical response to the problems and possibilities presented by the new age of responsive machines that the war had created. Many of the prominent figures in British cybernetics had been working outside their own fields for the war effort, allowing for interdisciplinary connections to be forged between, for example, machine, human and animal behaviours. Several of these individuals became members of the Ratio Club after the war, which was a discussion and networking group founded by W. Ross Ashby as a forum for the discussion of cybernetics. In 'The Origins of British Cybernetics: The Ratio Club', Owen Holland and Phil Husbands have described the 'coalescing of biological, engineering, and mathematical frameworks' within the club, noting the various roles that members had held in the war effort, from researching radar and gunnery to designing automatic control systems. They also observed that:

> In Britain there was little explicit biological research carried out as part of the war effort, so most biologists were, following some training in electronics, drafted into the main thrust of scientific research on communications and radar. They became part of an army of technical 'wizards' who Churchill was to later acknowledge as being vital to the allies' victory.[4]

Researching increasingly systematized modes of weaponry and defence created a diversely interdisciplinary experience for these biologists, most of whom were 'naturally unconstrained and interdisciplinary thinkers' who had, during the war been exposed to 'more explicitly mechanistic and mathematical ways of conceiving systems than they were used to'.[5] Peter M. Asaro has noted that in the interwar years too, British psychologists had debated 'possible mechanisms that could achieve mental capacities'.[6] Holland and Husbands offer a thorough account of how early British cybernetics dovetailed with psychiatry, converging upon the ambitious project of the artificial brain. They write:

> To these biologists a radar set could be thought of as a kind of artificial sense organ, and they began to see how the theoretical framework associated with it – which focused on how to best extract information from the signal – might be applied to better understanding natural senses such as vision.[7]

They emphasize that war technology was built into the roots of cybernetics, noting that 'several club members were deeply involved in the war-time development of early computers and their use in code cracking. This in turn brought them to ponder the possibility of building artificial brains inspired by real ones.'[8] Accounting for the overlapping interests across disciplines, they continue: 'Other engineers and theoreticians, working on such problems as automatic gun aiming alongside their biologist colleagues, began to see the importance of coordinated sensing and acting in intelligent adaptive behaviour, be it in a machine or animal.'[9]

The need for interdisciplinarity during the war created a new coalescence between researchers. Donald Mackay, a member of the Ratio Club, later recalled that the theoretical proposition of creating an artificial brain was very much part of the analogue/digital debate. His words on this subject are worth quoting in full:

> I found myself grappling with problems in the design of artificial sense organs for naval gun-directors and with the principles on which electronic circuits could be used to simulate situations in the external world so as to provide goal-directed guidance for ships, aircraft, missiles and the like. Later in the 1940's, when I was doing my Ph.D. work, there was much talk of the brain as a computer and of the early digital computers that were just making the headlines as 'electronic brains.' As an analogue computer man I felt strongly convinced that the brain, whatever it was, was not a digital computer. I didn't think it was an analogue computer either in the conventional sense. But this naturally rubbed under my skin the question: well, if it is not either of these, what kind of system is it? Is there any way of following through the kind of analysis that is appropriate to these artificial automata so as to understand better the kind of system the human brain is?[10]

What kind of system is it? Mackay's question ties together the various strands of process, behaviour and maintenance that formed the puzzle of the artificial brain. Considering the technologies that the Ratio Club members were working with, it is worth noting that as it stood, the 'thinking machine' was a rudimentary, unpredictable structure. As Cedric Price reminisced, visitors to Gordon Pask's digs were confronted with a thousand dripping mustard cans leaking water under the door: his computer.[11] He had built it in order to predict how many men would be working for the Royal Airforce in the year 2000.[12] Pask's other inventions trailed wires, constructed from war detritus and theatre lights. Likewise, Ashby's experimental machines were built from wired together scrap components, with dripping water troughs. These were in essence experiments towards the artificial brain: wet, with tendrils of wire nerves.

In Britain, the relationship between psychiatry and cybernetics was particularly notable, and figures such as Ashby, Pask and Grey Walter created a body of research and theory that pursued the reproduction of the human brain as machine. The human brain still stood as the most complicated and successful system known to man, and by extension, its replication represented the ultimate mastery of the living world. During the war, machines had been able to communicate with mankind, and with each other, and so this ambition to build a perfect artificial brain had moved out of the realm of science fiction, and into the realm of possibility. Machines had performed calculations and broken codes that were beyond the capacity of the human mind. In light of this, the cybernetic desire to model an artificial brain reflects a vital point of transition. It was clear to the first wave of cyberneticists that the progress taking place in computing meant that our relationship with machines had changed forever.

In *The Mangle of Practice: Time, Agency and Science*, Andrew Pickering writes compellingly of the philosophical ramifications of the surge of new machine technologies: 'the machine, as I conceive it, is the balance point, liminal between the human and nonhuman worlds (and liminal, too, between the worlds of science, technology and society).'[13] In Pickering's description, the poetry of the machine is precisely expressed, as is its place at the heart of the cybernetic project. This liminality was certainly one of the major themes for the many artists working with machines in the 1950s and 1960s, cybernetic and otherwise, including Ascott, Jean Tinguely, Nicolas Schöffer, György Kepes and Nam June Paik. Indeed, in the context of the visual arts, machine liminality opened out enticing possibilities of interactivity and performance.

There is an interesting thread that runs through the early cybernetic machines created by scientists, created by artists, as well as those produced through collaboration. Mechanical performativity was not only a pressing issue for artists working with machines – as Pickering has noted, early cybernetic machines often surprised their creators by 'behaving' in a manner they hadn't predicted, for example, self-correcting problems and thereby evolving their systems. Pickering describes this as 'performative ontology'.[14] He proposes that '[…] we need to move towards *ontology* and what I called the *performative idiom* – a decentred perspective that is concerned with agency – doing things in the world – and with the emergent interplay of human and material agency'.[15] In *The Cybernetic Brain: Sketches of Another Future*, he took the cybernetic 'mangle' of practice further, examining the interplay between the work of selected British cyberneticists and 1960s counterculture. He used his concept of performative ontologies to express the coexisting, interacting and adapting material and conceptual networks of the era. Pickering writes: 'We might thus think of cybernetics as staging for us a *performative epistemology*, directly engaging with its performative ontology – a vision of knowledge as *part* of performance rather than as an external controller of it.'[16] In Pickering's framework for cybernetics then, the investigator is drawn into the system, ceding control of the outcomes. There is no objective control, no analytical distance. This becomes impossible when machines are unpredictable, performing possibilities that extend beyond the intended boundaries of the intended experiment.

Ascott's analogue structures

Ascott began to create analogues in the period immediately after he had first come across cybernetic theory in the King's College library, and this, coupled with his direct experience of communication technologies in Fighter Control, prompted him to evolve an increasingly technologized theory of communication in the visual arts. This theory was partly expressed through Ascott's art practice, in part through his writing and finally, it was expressed in a diagram he produced in 1963 for the *Diagram-Boxes and Analogue Structures* exhibition catalogue.[17] It was during the brief spell that Ascott spent as a studio demonstrator at King's College after completing his degree and before taking up the post of head of Foundation Studies at Ealing College of Art, that he 'discovered' cybernetics. Ascott's exposure to cybernetic ideas proved to be a profound and transformative moment, the start of a career that would be defined by his deep connection to the cybernetic model of communication.

Ascott's own account of this important phase of his career focuses on the moment, when browsing the stacks in the library, that his hand alighted upon F.H. George's *Automation, Cybernetics and Society*.[18] This was recounted by Pethick in her short article *Degree Zero* as Ascott's 'Eureka moment', when he 'came in contact' with cybernetic theories of animal and machine.[19] The notion came from Ascott himself, who latterly described it as a 'Eureka experience – a visionary flash of insight in which I saw something whole, complete, and entire'.[20] Ascott remembers this moment with fondness, describing the way in which we browse libraries, the chance encounters that lead us to new discoveries.[21] However, this 'Eureka' moment was not solely prompted by the novelty of cybernetics, but rather from finding a theory that encapsulated Ascott's developing interests in human communication and technologies.

Despite this moment of recognition and inspiration, it is important to note that Ascott's work leading up to this point already used principles of interactivity and change, and from this we can deduce the dual influences of an art training focused upon process in art and nature, alongside Ascott's earlier experience of communication technologies in Fighter Control. Cybernetic theories grew in the post-war period as a result of the new interactive technologies of World War II and the technical, moral and philosophical questions they posed for mankind's mechanical future. In this way Ascott's exposure to cybernetics was a direct one, in that he worked with many of the technologies that had provoked the field of cybernetics. These dual influences created the grounds for Ascott's lifelong interest in systems, change and process – in cybernetics as a living, vibrant model of interaction. Indeed, the philosophies of artistic practice derived from systematized biology that was embedded into the pedagogy at King's College had several points of overlap with early cybernetics. F.H. George wrote in the introduction to his 1962 book *The Brain as Computer* that:

> The title of the book [...] is intended to convey something of the methodology involved; the idea is to regard the brain itself as if it were a computer-type control system, in the belief that by so doing we are making explicit what for some time has been implicit in the biological sciences.[22]

Cybernetics expanded the work of systems biologists in response to the new knowledge, both biological and technological, generated during World War II. Ascott began to articulate a vision for the visual arts in response to this new age of both human and mechanical communication.

Video Roget: the cybernetics of abstraction

By 1962, Ascott had incorporated direct references to cybernetic themes such as feedback systems into his work, while simultaneously exploring the relationships between words and shapes in analogue structures such as *Video Roget* (Plate III). Created in 1962, the influence that Pasmore's organicism had for Ascott is still very much evident in the series of forms in wood, and the title, in its reference to the thesaurus, connects it to the Basic Design presentation of a 'grammar' of form. Each wooden symbol explored the vital interest Ascott had developed as a student in nested language: umbrella/shelter or bottle/container, single forms with multiple, connected possible meanings. In the centre of *Video Roget*, in common with Ascott's other analogues, there is a thin strip of Perspex. It is green, inscribed with two parallel black lines, each divided at irregular intervals by three short vertical lines. In front of the Perspex, a single square of Plexiglas with a vertical red line can be moved, by the participant, from side to side across the width of the analogue. The red line could be aligned with the forms immediately above and below the diagrammatic central strip of Perspex, so that the viewer might build relationships between the forms.

Sets of related forms are arranged in rows. They play upon processes of growth and change in nature – rounded shapes that are divided like cells, stemmed like fruit or leaves, cracked open, sprouting, segmented or split. The interrelationship between the forms is ambiguous but still compelling; there is something evolutionary about the way in which a wedge-shaped intersection into a sphere anticipates the fuller spiral adjacent to it. Ascott plays upon the beauty of the limited number of simple forms and structures that are the basic language of every complex form in our world. Branching lines, spirals and crystalline structures are part of a lexicon of growth in nature, and the means of creating imagery of such structures of the natural world had advanced considerably in the post-war years.[23] In this way, the permutations of biological abstraction that took place in the post-war era differed from those in the interwar years, since ever-more complex microscopic structures were being made visible.

Rudolf Arnheim's philosophy of form was an essential element of the debates that still reverberated in British art a decade after Hamilton's *On Growth and Form* exhibition.[24] When considering Ascott's interest in form as process, Arnheim's later work has some relevance. Carlos Augusto Moreira da Nóbrega and Maria Luiza P. Guimarães Fragoso have noted Arnheim's 1994 essay 'The Completeness of Physical and Artistic Form', in which Arnheim comments on the complication of form in the realm of human perception; how in fact, 'a new understanding of form could be delivered from a world-view.[25] They quote Arnheim:

> form is an abstraction [...] which combines bodies and forces. In this more complex view, bodies in and by themselves remain as static as before, but now they are seen as inhabited by forces, forces that move them and let them act on other bodies. [...] This more unified notion abandons matter as a separate concept and leaves organized energy as the only and sufficient substrate of the universe. What looked like bodies is nothing but an agglomeration of forces.[26]

Shape, form, the organized body, then, are all shaped by 'forces' – form is in itself an abstraction that held decreasing relevance for a generation of artists that were living through the birth of the systems age. During Ascott's education at King's College, both Hamilton and Pasmore had been intensely occupied with the creative possibilities of biological form as described in the previous chapter. The Basic Design curriculum enclosed biological form within its overall grammar of form, or grammar of possibility; it is unsurprising then, that Ascott began to work with the principle of change as an extension of this teaching philosophy. These biological symbols became a site for interchange and growth for Ascott, and it is worth noting that advances in scientific imaging revealed just how systematized the natural world is, that living form is mechanical. That is to say, every organism is created from a series of interacting elements that together sustain the success of a system.

When Ascott began to read cybernetic theory, he discovered that this idea was articulated clearly. Early in *Design for a Brain*, Ashby wrote that: '[…] the free-living organism and its environment, taken together, may be represented with sufficient accuracy by a set of variables that forms a state-determined system.'[27] If we read *Video Roget* in relation to this, it is evident that Ascott was exploring, in abstract form, the overlapping territories of mechanical biological and cybernetics. Ascott placed abstract biological form within a network of possibility, emphasizing the notion of nature as change and emphasizing change cycles of birth, growth and gestation. The subtle grids of lines created exchanges between forms, a semiotic game.

During Ascott's studies Hamilton had been preoccupied with his meticulous translation of the ninety-four documents that made up Duchamp's green box.[28] Hamilton had collaborated with George Heard Hamilton, who had performed the translations from French to English. As Paul Thirskell has described, Richard Hamilton saw his role as that of a 'monolingual translator', who could 'capture the underlying concept'.[29] He spent a considerable amount of time negotiating the dissonance between French and English, and between word and symbol in order to construct a set of meanings that he considered 'true' to Duchamp's original. Ascott and the other students at King's College were comprehensively familiar with Hamilton's project. There are some parallels between Ascott's *Video Roget* and Hamilton's reconstruction of Duchamp, each with an intricate play upon the treachery and possibility of information, language and symbolism. While the two works of art are very different, each worked through the application of a textual framework of interpretation to an art object. Hamilton's all-encompassing task also exposed the dissonance between object and information, which positions both projects within the context of systems of communication. At the same time, it is worth reiterating the increasing currency of the term 'information', which grew in prevalence in the visual arts against the backdrop of its use in early computing. Jack Burnham later articulated this within a cybernetic context in 1968 in the article 'Systems Esthetics', writing that 'For systems, information, in whatever form conveyed, becomes a viable esthetic consideration'.[30]

The semiotic games around *Video Roget* become more complex when we examine its reproduction in the Molton Gallery catalogue. Ascott's catalogue included a 'thesaurus' diagrammatic overlay printed on a transparent page of tracing paper that aligned with the catalogue photograph of *Video Roget*, mapping out Ascott's definitions for the formal elements of the work (Figure 2.2). This overlaid diagram corresponded with the abstract shapes depicted on the page underneath. In this diagram, the top row of shapes was labelled 'meta-forms for organic growth'; Ascott used his term 'metaform' to indicate that form held multiple possibilities, that form was potential.

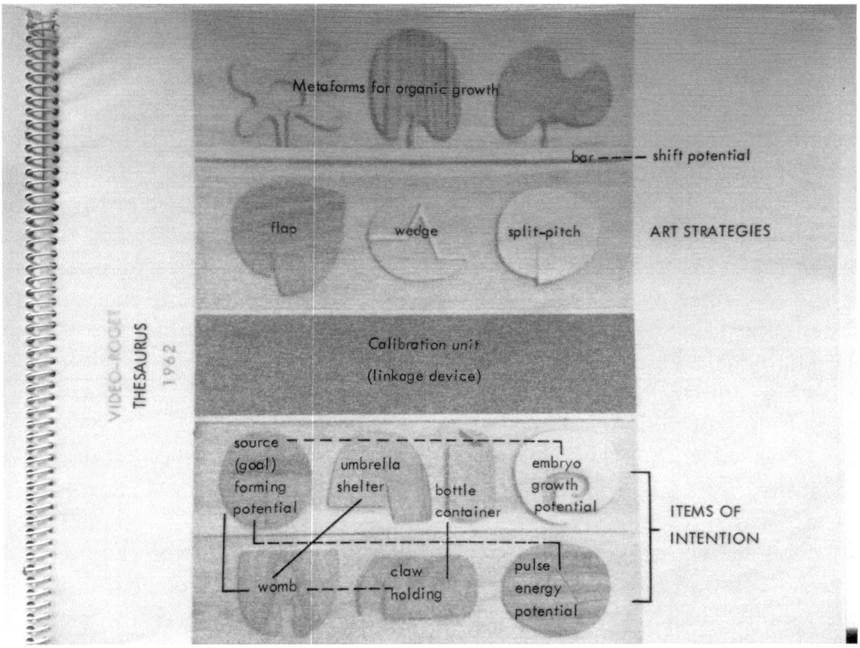

Figure 2.2 Roy Ascott (1962) *Video Roget.* Analogue structure: Plexiglas, wood and glass, 50 × 35 in. (from Diagram-Boxes and Analogue Structures catalogue, Molton Gallery, London 1963)

The second row down was labelled 'art strategies', denoting three ways of dividing or bisecting a single form and playing upon the fact that the artist 'intervenes', both materially and conceptually. The term 'strategy' implies both game play and objective planning, far removed from Pasmore's more intuitive approach to simple form.[31] The central strip of Perspex was marked as a 'Calibration Unit (Linkage Device)'. This is Ascott's earliest formal reference to the concept of calibration, which would later prove to be one of the core strategies of his Groundcourse pedagogy. Here, Ascott used the term to express the fact that the viewer could, through moving the elements at the centre of the analogue, calibrate it against their own intentions. Other applications of the term in military and technical context will be discussed more fully later in this book.[32] For now, I will note that during and after World War II, the RAF used the term 'Calibration Unit' for anti-aircraft squadrons, and a small number of analogue devices such as flight calculators were described as 'calibrators' too.

The bottom two rows of the analogue were labelled 'Items of Intention'. These were two rows of 'metaforms' that Ascott had individually labelled to indicate the multiplicity of meaning for each form, such as bottle/container or umbrella/shelter. Shanken argued that in *Video Roget* '[...] meaning was contingent on the flow of information between the artist, the semantic systems that govern the reception of works of art, and the actual responses of viewers'.[33] The nested language of Ascott's metaforms created a network of possibility for the viewer, according to the path of meaning they created around and between forms. In this way, *Video Roget* played upon the instability of visual language to create a positive and essentially objectivist vision of how meaning is constructed, and reconstructed, around the work of art. In the many examples of interactivity in late British abstraction, there was a fundamental shift in which meaning was located not in the art object itself, but in the discourse around the object. Ascott's understanding of the discourse a work of art might provoke was becoming increasingly systematized; his work enclosed the viewer in a process of interaction with the analogues. As Ascott immersed himself in this language of biological process – 'embryo/growth/potential' – he came to the conclusion that form, in itself, had behaviour.[34]

Concealed within the modest pages of the Molton Gallery catalogue is one if the least-recognized manifestos of modern art. Ascott had put together a chart which resembled an electrical circuit, creating a set of complex relationships between art, society, cybernetics and communication (Figures 2.3 and 2.4). In one box at the bottom right Ascott wrote, somewhat ambiguously, that: 'This thesaurus is a statement of my intention to use any assembly

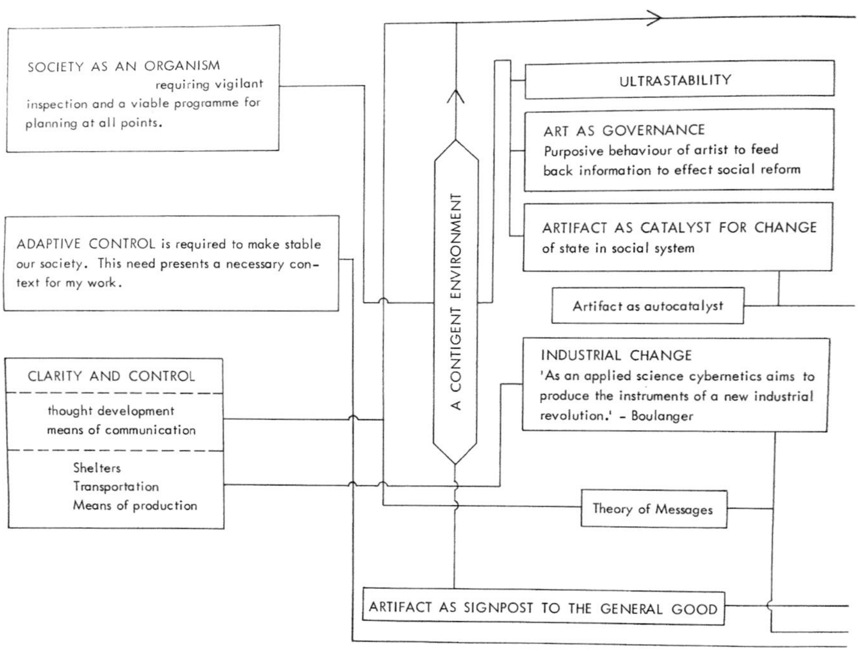

Figure 2.3 Roy Ascott (1962) *Left Page of Diagram from the Molton Catalogue (A Cybernetic Manifesto)*. Illustration designed by Noel Forster

of diagrammatic and iconographic forms within a given construct as seems necessary.' The diagram operated as a manifesto for a situated cybernetic art practice, and after more than fifty years, it is still the most comprehensive manifesto for cybernetic art. This double-page diagram (shown here in two halves) mapped out on the left an explicitly social model for art practice and on the right, the systems philosophy behind Ascott's practice, the two sides integrated by connecting lines between the boxes that contained each proposition. It is a flow of connected ideas, each influencing, and influenced by, the next. On each side of the diagram, there is one single proposition displayed vertically – on the left the term, 'A Contingent Environment' and on the right, 'Work Pattern for Diagram Boxes and Analogue Structures'.

On the top left, Ascott wrote 'SOCIETY AS ORGANISM, requiring vigilant inspection and a viable programme for planning at all points'. This reflects the cybernetic principle that all organisms are systems, in constant interaction with their environment. Directly opposite this on the right-hand side of the diagram, is the term 'ULTRASTABILITY'; a word taken from W. Ross Ashby's *Design for a Brain*, in which he describes how adaptation is necessary to the maintenance of an 'ultrastable' system.[35] Ascott was referencing Ashby's elegant proof that adaptive behaviour, in both machine and organism, is directed towards the maintenance of stability; or homeostasis. The diagram links 'ultrastability' with two other phrases; the interlinked

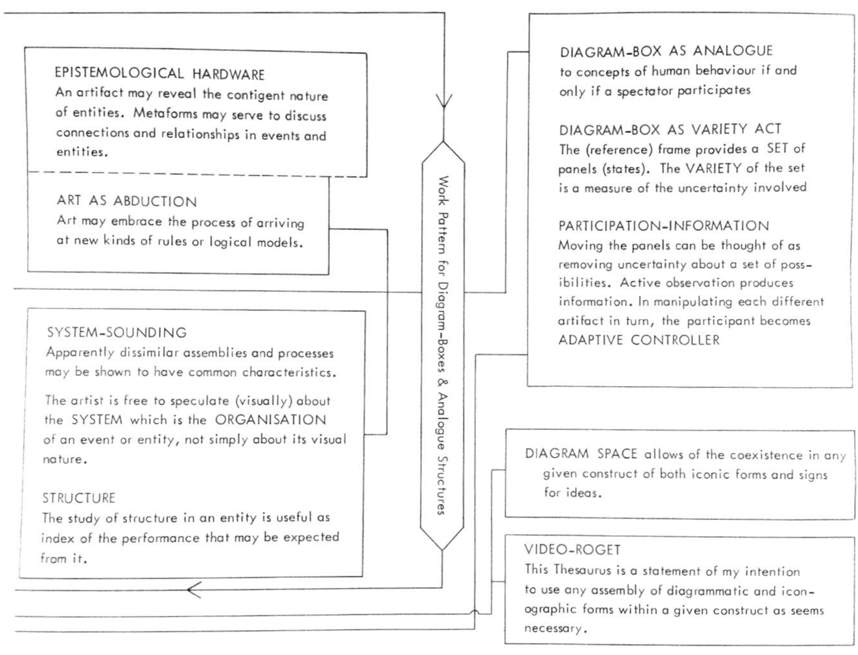

Figure 2.4 Roy Ascott (1962) *Right page of diagram from the Molton Gallery Catalogue (A Cybernetic Manifesto)*. Illustration designed by Noel Forster

boxes below contain the instructions 'ART AS GOVERNANCE Purposive behaviour of artist to feed back information to effect social reform' and 'ARTIFACT AS CATALYST FOR CHANGE of state in social system'. In this way, Ascott builds a picture of society as a homeostatic system, with the work of art a 'catalyst for change'. Ascott places art as a potential disruption that might make a system 'adapt' in order to maintain its equilibrium. The artist has the power to influence social change; operating between the audience and the work of art, the artist 'feeds back' information that might be a catalyst for 'social reform'.

On the bottom right, the term 'theory of messages' is boxed in, wired to the circuit of Ascott's diagram. Ascott had been profoundly influenced by C.E. Shannon's seminal article 'A Mathematical Theory of Communication' which had first been published in the *Bell System Technical Journal* in 1958.[36] This article established the field of information theory, examining the ways in which a piece of information (a message) flowed from source to transmitter, transmitter to receiver, receiver to destination. It included a diagram of boxes, connected by lines entitled 'a schematic diagram of a general communication system'. Vitally, it displayed information as a flow that moved between units, and might be changed by its journey. Ascott drew together Ashby's thesis of homeostasis with Shannon's communication theory.

Ascott intended the diagram as a key to his model of practice for the body of work in this exhibition, but it was in essence a fully realized theory of cybernetic art and its production, reception and change in converging environments and contexts. It was in essence the same ideological approach that Ascott would use when designing Groundcourse later the same year. This work of art and its subsequent presentation for the Molton Gallery catalogue serves as a physical representation of Ascott's transition from the abstraction based on systems biology of Basic Design to a cybernetic model for art practice. It is an elegant manifesto for cybernetic art: a diagrammatic visual representation of the converging ideologies which contributed to new forms of visual art practices in the 1960s.

On the right-hand page, Ascott included more specific references to his own working practice. He gave a guide to understanding the work, explaining that 'diagram-box as analogue to concepts of human behaviour if and only if a spectator participates', sketching the analogous relationship between participant and interactive work of art and enforcing that the analogous element was 'human behaviour'. Ascott was describing participation as a mode of behaviour, in which the decisions made by the participant were physically measured by the analogue work of art. The use of the term 'behaviour' is an example of the overlapping language of science, engineering, psychology and art that was simultaneous to, and often provoked by, early cybernetics. For Ashby, for example, both organisms and machines exhibited 'behaviour', and in this way behaviour was a measurable element of systems analysis.

Further, Ashby was hoping to demonstrate, in *Design for a Brain*, that by replicating the processes of the human mind '[...] a machine's behaviour may be made as adaptive as we please, and that the method may be capable of explaining even the adaptiveness of Man'.[37] In early cybernetics then, the mechanical and the biological were fused at the root and particularly in Britain, and as noted in the introduction to this book, the discussion of future technologies was intimately linked to psychiatry.[38]

There is an amorphous relationship between the fields of behaviourism and cybernetics, and Ascott had also read the work of B.F. Skinner, whose work before, during and after the war defined the development of behaviourism.[39] In this way, Ascott's use of the term behaviour brought together tactics of social and psychological control, tactics of systems control and the fusion of systems both living and engineered. Ascott's own application of the term to both the formal elements of his work ('form has behaviour') and to the decisions of viewer-participants (when 'the spectator participates') indicates the emerging importance of the principle of control in his work. For while 'behaviour' was an observable systems phenomenon, it was also something to be modified, provoked or changed. Ashby intervened in the function of the homeostat so that it might adapt and try to regain its ultrastability, Skinner trained his laboratory rats through a variety of disturbances and Ascott himself used his analogues as 'catalysts for change'.[40]

Ascott mentioned control explicitly three times in this diagram – he makes two references to 'adaptive control', firstly as necessary to 'make stable' our society, secondly placing the participant in the role of 'adaptive controller'. There is an explicit link between his concepts of stability and of control; as I have accounted, stability in this context relates to homeostasis as the maintenance of a stable state in a system. The term 'adaptive controller' then, highlights an important power relationship in that an artist or participant might disrupt or agitate a system through either the production of 'artefacts' or through interacting with those artefacts. On one level, we might consider the participant who takes the position of 'adaptive controller' by interacting with Ascott's analogues as an actor in a performance of systems behaviour. However, in Ascott's vision, this creates ripples that might affect social or political systems in turn. One system converges on the next.

A diagrammatical turn: communication technologies and conceptualism

The many layers of *Video Roget* together operate as a work of art, a manifesto and a game. It represents a moment of cohesion in Ascott's early career, in which his increasingly interactive use of abstract construction fused with a set of cybernetic principles. It is a physical link between the mechanized modes of biological abstraction of the 1950s and British cybernetic art; they shared a point of origin in the principle of living form as process. It is also worth noting that Ascott's play upon semiotics and the processes and problems produced by diagrammatizing meaning was representative of what was emerging as an important facet of art-making, one which would be manifest in different strains of conceptual art throughout the 1960s and 1970s.

As Cristina Iuli has noted in her essay on the place of cybernetics in 1960s art theory and practice, '[...] by the mid-1950s the concepts of *information, communication,* and *systems* had grown common in the vocabulary of the hard sciences and were disseminated across social and aesthetic practices and discourses'.[41] Iuli argues that these associated ideas and practices amounted to an epistemological revolution that 'generated a shift in the concept of art and in the notion of the artwork from object to process, from static to performative, and from closed to open system'.[42] Cybernetic art has long stood as a kind of peripheral and specialized footnote to the great and subversive explosion of visual arts practice in the 1960s, confined by specialism, by the fact that the majority of work was lost, destroyed or unsalvageable, and also due to its complexity.[43] This is despite the fact that in essence it represents a pure core of the systems ideology that is a uniting element in the disparate practices of the post-war years. The concepts and associated vocabulary that shaped an age – *systems*, *information*, *communication*, *feedback* – were broadly disseminated, but the fact that these terms were cybernetic was not broadly understood. That misunderstanding has endured within the canon of art historical discourses.

The most obvious context for Ascott's *Video Roget* 'thesaurus' is the emerging field of cybernetic art, in which explanatory keys, descriptions and diagrams were used frequently. This evolved for practical reasons – artists that were using complex theory and engineered design in their practice created a distinct obstacle for their audiences. Both the technologies and the ideologies of cybernetics were complex and not broadly accessible. While the cyborg haunted the twentieth century as an emblem of a future lived through (or conquered by) technology, the science, mathematics and engineering that made the cyborg possibility rather than dream was a specialized realm. When in 1968, Jasia Reichardt's *Cybernetic Serendipity* exhibition at the ICA moved cybernetic art into the spotlight, the accompanying catalogue began with a glossary of terminology – that same terminology that had fascinated and beguiled Ascott nine years earlier.[44] Even after the decade that had passed, cybernetic terminology such as 'feedback', 'black box', 'analogue' and 'digital' needed to be defined, a common issue in any art that applies, or engages with, contemporary science.

The distinctly systemized turn in artists' writings in the 1960s through to the mid 1970s took place against the advent of systemized technology that in turn created our own network age. The creation or application of communication 'systems' in all fields of art practice should be understood as an outcome of technological growth; it was Shannon's 'A Mathematical Theory of Communication' that first articulated the idea of information flow, an essay provoked by, and intended to address, the complicated exchanges of information in early computing. Donna Haraway acknowledged that Katherine N. Hayles charted the translation of Shannon's theory from a theory developed 'for packing the largest number of signals on a transmission line for the Bell Telephone Company' and then 'extended to cover communication acts in general, including those directed by the codes of bodily semiosis or molecular biology'.[45] Haraway also noted the significance of Shannon's communication model to posthuman conceptions of the body, as well as how communication theory moved from the technological to the physiological, the psychological and to the arts.[46] The appropriation and new applications of Shannon's work took place swiftly and progressively; by 1953, Bar-Hillel and Carnap had written 'An Outline of a Theory of Semantic Information', arguing that it would 'serve as a better approximation for some future explanation of a psychological concept of information than the concept dealt with in communication theory'.[47] In this moment communication theory diverged to follow two paths, one furthering Shannon's research into information transmissions and the other following a philosophical discussion of semantic information.[48] Despite the divergence of the two areas of research, the principle of a flow, or process, of information retained its original framework as delineated by Shannon.

For Ascott, Shannon's article worked as a framework for human communication, due to the malleability of the core concept of communication as 'information' that passes from one source to another, potentially experiencing disturbances or alterations along the way. Haraway also noted: '"Information" generating and processing systems [...] are postmodern objects, embedded in a theory of internally differentiated signifiers and remote from doctrines of representation as mimesis.'[49] In this model of communication, the art object is never complete, remaining caught in the process of information flow with all its potential fluctuations of content. Shannon wrote: 'Frequently the messages have meaning; that is they refer to or are correlated according to some system with certain physical or conceptual entities.'[50] In this way, the engineering notion of meaning – as something that refers beyond the boundary of the message – correlates quite exactly with the notion of the art object as a carrier of meaning. It is easy to see why Ascott and other artists immediately noted the relevance of this article to communication in the visual arts, integrating it with his developing framework for a cybernetic system of art making.

Shannon's theory was frequently illustrated with flow diagrams and the conceptual malleability of his theory is immediately evident; he describes phases of information flow with a series of interlinked boxes with the potential disruption of a 'noise source'. At the time of Shannon's writing, the term 'information' was rapidly accruing new applications, beyond its original meaning of a body of knowledge about a particular topic. It had been appropriated by electrical engineers in the 1930s to describe television signals, and after World War II it was used in relation to punch card data. Finally, by 1966, the so-called 'information revolution' became a common term as computing advanced. It is no coincidence then that the electrical applications of the term 'information' and the advancing theoretical discourse around visual arts as 'communication' occur in the corresponding timeframe. Electrical information is disembodied; it transmits as a signal from a source to a destination, sometimes altered along the way. In the context of the transformations from object to process in the visual arts that took place following World War II, there is a deeply rooted connection between the disembodiment of electrical information in technology and the disembodiment of conceptual 'information' in the arts. In Shannon's model of communication, information is not fixed – its journey alters its form. If we apply this to the art object, then the variables of viewing and transmission, reception, discussion, digression and even transmutation of meaning form an information flow that is both complex, and ultimately unstable. Looking at this trend retrospectively, it is compelling that the art strategy of diagrammatizing emerged in conjunction with – or in response to – this instability of meaning.

Like Shannon's flow of disembodied radio signals, the flow of 'information' in organisms is vulnerable to interference and subject to transformation between source and destination. In reviewing the overlapping territories in which information flow became a focal ideology following Shannon's seminal publication, it is evident that across disciplines, communication was subject to the same issue of transformation through process. That is to say, there is no guarantee that a message of any kind will reach its destination intact, because the process of communication itself is, by nature, often transformative. As Gilbert Simondon described:

> Mental systems influence each other during invention in the same way as different dynamisms of technical object influence each other in material functioning. The unity of the associated milieu of a technical object has an analogue in the unity of a living thing.[51]

The cybernetic principle behind Simondon's words, that living and engineered systems are analogous in terms of behaviour, was of increasing importance to Ascott's practice in the early 1960s. There is a dizzying quality to the concept of information flow – a unifying energy connecting discrete systems in a web of interaction, the messages taking many forms. With this in mind, it is easy to understand why cybernetics was to become a curious but ingrained facet of counterculture, contributing theories, practices and concepts to psychedelic art and music.[52]

Booths and black boxes: analogues at the Centre D'Art Cybernétique

In 1964, Ascott had a solo exhibition at the Centre D'Art Cybernétique in Paris, a gallery dedicated to cybernetic art that had been founded by

Suzanne de Coninck in 1962. A large number of the works of art that Ascott displayed have since been lost, although they were recorded in a series of installation photographs. Amongst the *Change Paintings* and *Analogues* on display, there were a series of works that were constructed as walled-in shelves, giving the appearance of booths. This included *Homage to C.E. Shannon* (Figure 2.5), a triple booth with the two right-hand interior spaces painted black. On the left was an upturned tin funnel, hovering above five upturned tin cups, on the right a shelf with round holes containing upright wooden objects that resembled game playing pieces. The centre space was empty, a depthless black vacuum.

This analogue is both formal and playful, using a combination of domestic objects and carved wooden elements. There is suggestive tension about this analogue; it is like a stage set carefully arranged before a performance, awaiting inevitable disruption. The central booth in particular invites intervention through the movement of elements from the left or the right. The set-up of these black booths is reminiscent of phone cubicles or post office booths, spaces of solitary communication. This stage set of objects invites considered interaction; the careful arrangement of forms awaits disruptions. Ascott's early analogues are atmospheric, concentrated and still playful. Ascott recalls that this sculpture, created for the newly founded Centre D'Art Cybernétique in Paris, was also a creative response to the information theories of C.E. Shannon. As previously noted, Shannon's theory of messages described how information flowed from source to transmitter, transmitter to receiver, receiver to destination.[53] The central space can therefore be read as an abstract space of communication, in which 'messages' might be transmitted and received. In this light, the cups and funnel employ a playful metaphor for the flow of information as an entity that might be directed (funnelled), transported (contained in cups) and transmitted (moved into the central space of 'transmission').

Ascott was also interested in the suggestive poetry of the black box, a concept that had been most fully theorized by W. Ross Ashby, an author Ascott had read directly after F.H. George.[54] Ashby explained that the '[...]

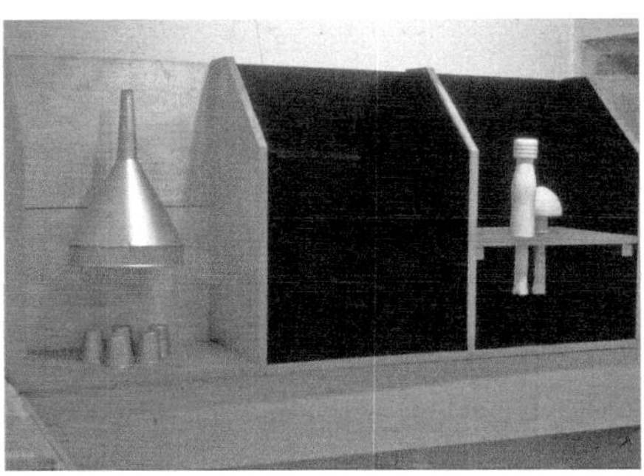

Figure 2.5 Roy Ascott (1964) *Homage to C.E. Shannon*. Centre D'Art Cybernétique, Paris

problem of the Black Box arose in electrical engineering. The engineer is given a sealed box that has terminals for input, to which he may bring any voltages, shocks, or other disturbances, he pleases, and terminals for output from which he may observe what he can.'[55] As an exercise, this neatly encapsulated what was the greatest issue of contemporary technologies – the unpredictability, and unknowability, of machines. Scientists including Ashby were frequently surprised by the behaviour of mechanical systems, their ability to self-adapt, the frequent discoveries that they made as by-products of an experiment. This new condition of working in the sciences was what prompted Andrew Pickering to so usefully suggest a 'performative ontology', a mode of knowledge pursuit that takes in the quasi-theatrical 'performance' of possibility that modern mechanical engineering had produced.[56]

Ashby suggested that we are, in our daily lives, surrounded by black boxes – mechanisms, or other objects in the world, whose internal workings are not visible from the outside. As the mechanical engineering students were discovering, the black box could be best understood through operation and observation, rather than by dissecting it. This simple point went against the grain of empiricism, and perhaps of modernity itself – rather than breaking open the black box, you must work with it to gain understanding, working on the premise that it will still remain unknowable. Pickering suggests that this mode of working – accepting unknowability as a preposition – resists modernity, writing that 'Modernity is defined by projects of domination, then cybernetics is marked by a symmetric accommodation to the ultimately uncontrollable'.[57] To illustrate the idea of an everyday black box, Ashby describes a child opening a door using its handle, unable to see the internal mechanism but learning its potential through opening the door. There were, he said, black boxes everywhere. He describes the way in which a ship engineer both observes its operation and enacts change upon its operation:

> We now see the experimenter much like the engineer in a ship, who sits before a set of levers and telegraphs by which he may act on the engines, and who can observe the results on a row of dials [...]. The representation [...] is [...] capable of representing the great majority of natural systems, even if biological or economic.[58]

In this way the black box problem frames the complexity of mechanisms, their ability to self-regulate and the philosophical shift that scientists had to make from isolating mechanisms to 'performing' their functions.

The duality of the concept of performative ontologies offers a framework with which to assess the ways in which visual arts engaged with cybernetics. One of the curious and wonderful features of the cybernetic field is the way in which a purely scientific ontology twinned itself with a parallel shift in philosophy; how, for example, the practical problem of the black box is mirrored by the black box as a 'performance' of the problem of objecthood.

Norbert Wiener might have treated the interdisciplinary appropriation of cybernetic theory with some cynicism, but one British practitioner had from the outset a love of the peculiar and fantastic interactions of an interdisciplinary cybernetic world.[59] Gordon Pask, who was a close friend of Ascott's, wrote in 1962 that:

> Now, we are self-organizing systems and we wander around in a world which is full of wonderful black boxes, Dr. Ashby's black boxes. Some of them are turtles; some are turtledoves; some are mockingbirds; some of them go 'POOP!' and some of them go 'POP!'; some are computers; this sort of thing.[60]

In this arrestingly surreal description of modern mechanics, Pask playfully references Grey Walter's mechanical tortoises, which had so amused visitors to the Festival of Britain in 1951.[61] There is a sense of the magician's performance in this description; the turtledove emerging from the deep darkness of the magician's hat. Ascott was familiar with the writing of both Pask and Ashby by 1961 and by 1964, had fully immersed himself in cybernetic theory.

Drawing Shannon and Ashby together then, *Homage to C.E. Shannon* played upon visibility and the unknown, the liquid darkness of the booths representative of the performance of communication that characterized the new age of technology. The booth format was also a convenient space of private interaction; like the booths at a bank used for filling out cheques, or making telephone calls in hotels, they bring to mind the collision of public and private interaction. They frame the space as a zone of individual communication in which the viewer faces the wall, reaches into the enclosed space, blocks it from public view with his or her body. This tension between the public space of the gallery and the private, transaction-like interaction with the analogue recalls the Fighter Control bunker, where individual offers worked at a solitary interface in an open, collectively functioning environment. In this sense, the recurrence of the booth as integral to modern architecture is suggestive of the need to delineate privacy in the shared spaces of modern life, whether commercial, military or personal. Public telephones, for example, were often wall-mounted in hooded booths, or in shelved booths in hotels, in a manner that created an illusion of privacy in open public space. A work called *Kiosk*, now lost, was also on display in Paris; Ascott described its manipulation as '[...] in the form of goal-less 'measurement' and sifting through papers in a built-in drawer that were visual comments on the structure that contained them'.[62] This lost work also played upon the privatization of space for an individual function, such as the kiosk of an office clerk. *Kiosk* was evidently a self-referential hermetic system, in which messages and meaning were analogic to shape and form.

Like other contemporary sculpture, the analogues disrupted the exchange between viewer and object, in this case through inviting touch and movement. Robert Morris's Green Gallery installation was also in 1964 – there are some parallels between the formal approaches taken by the minimalist artist to the object–viewer relationship, and those in late British abstraction. Both subverted the idea that the gallery was a space of display, treating it instead as a space of interaction and exchange. It is certainly a point of transatlantic comparison, given that Morris's exhibition played upon the formal boundary between sculpture and space in a manner that Rosalind Krauss famously characterized as sculpture that had '[...] entered the full condition of its inverse logic and had become pure negativity: the combination of exclusions'. She continued that it had '[...] ceased being a positivity, and was now the category that resulted from the addition of the *not-landscape* to the *not-architecture*'.[63]

The *Change Paintings* in particular resist traditional categorization; installed into at the Centre D'Art Cybernétique, they identify more as objects than as either painting or sculpture. At the time, Ascott did not characterize this work as sculpture, considering himself simply as an artist, working as he did with wall-mounted works that had gradually extended forwards into the gallery space from the two-dimensional *Change Paintings* to the three-dimensional *Analogues*.[64] This was true of several of his contemporaries, including his teacher Pasmore, who gradually evolved towards sculpture through construction, opening out from the two-dimensional plane as a means of disrupting the formal viewing conditions of abstract art. Despite the fact Ascott did not consider these works sculpture they were resolutely sculptural. The booths employed in the Paris *Analogues* had some relevance to Krauss's territory of resistance to categorization, creating architectural spaces as they did. However, Ascott's work and much late British abstraction and construction retained a built or constructive aesthetic, that made it appear sculptural. The analogues, as interactive works, were of the hand too. Ascott wanted the movement of the pieces by hand to provoke recognition that the work of art was something that changed through interaction, that change being both cerebral and physical.

In another installation shot from the Centre D'Art Cybernétique (Figure 2.6), there is partial sight of an analogue on the far left, which had pegs of roughly humanoid form – upright, rounded, the suggestion of heads. There is also a circular form with segments of dark wood that resembles a rotating fan, set on the panel parallel to the wall. Such circular motifs were present in other analogues of the period too; they are reminiscent of air vents or of plane propellers, the straight lines of the segments contrasting with the rounded shapes of the pegs. The blend of biological and technological motifs demonstrates how neatly Ascott's early preoccupation with mechanical biology dovetailed with the systems biology that was so integral to early cybernetics. At the centre of Figure 2.6, another analogue is of dark oval form, with a pale central panel that has a bulbous outline. At its centre, two precisely intersecting toothed panels meet horizontally. These 'teeth' are in fact a rhythmic set of jagged peaks that resemble the regular pulses

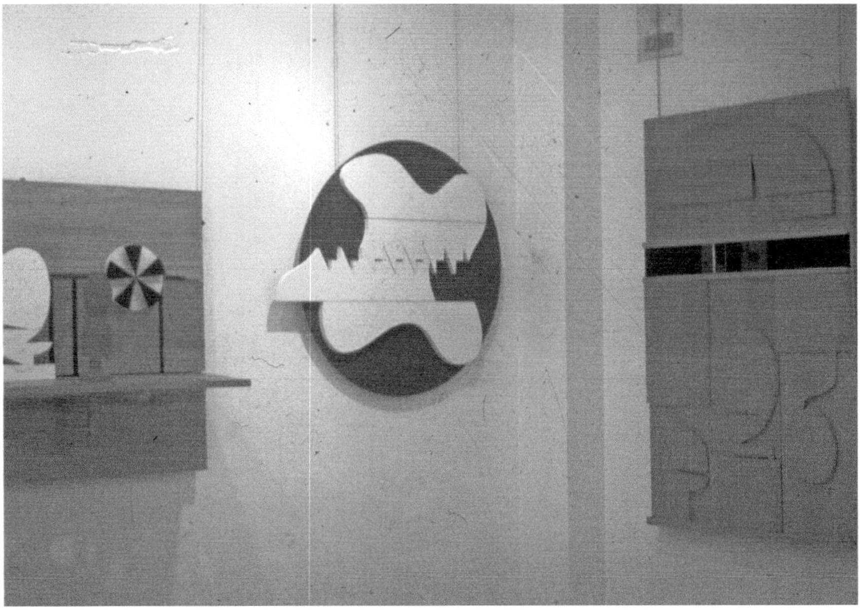

Figure 2.6 Roy Ascott (1964) *Installation Shot of Solo Exhibition*, Centre D'Art Cybernétique, Paris

of a radar reading. In Ascott's early analogues, signal was frequently transformed into visual code in wood or plastic.

The third analogue in Figure 2.6 is *Shelter* (1963), a work of art that spent 40 years stored in a Canadian barn, before being recovered, restored and exhibited in 2013 by the Institute of Contemporary Art in Winnipeg.[65] *Shelter* has an arching umbrella form in the top third, a 'metaform' that Ascott used to symbolize protection. Underneath a blue Plexiglas strip, there are three horizontal motifs, each resembling seeds pushing below or above the flat surface of the earth. The rounded seeds are each cleft, unfurling outwards. *Shelter* has formal similarities with two other analogues that were also restored in Canada – *Analogue Table* (Figure 2.7) and *Items of Intention* (Figure 2.8).

Analogue Table brings together the precise metaforms that Ascott had been employing for two years, along with a form of 'drawing' with a jig-saw to make fluid wave-energy patterns and other abstract marks, a process that he explored in a more focused way in the mid to late 1960s.[66] In off-white painted wood and jade green Perspex, it echoes the clean, modern schemes of contemporaneous interior design. The free-drawn cuts in the wood have a sense of humming energy, of motion and process.

The analogues are interstitial, hovering between biologically entrenched abstract symbolism (form in process) and the representation of blips, pulses

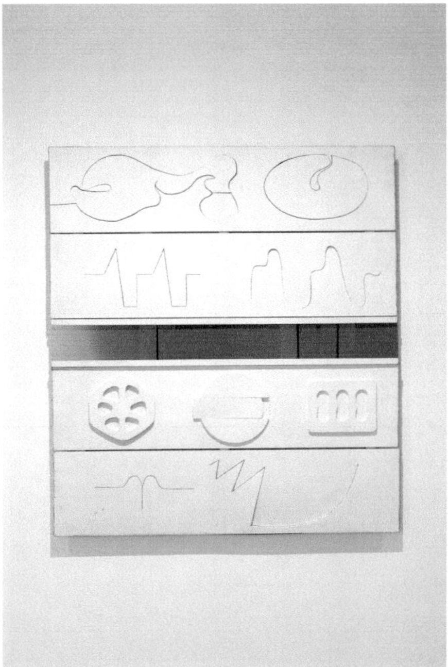

Figure 2.7 Roy Ascott (1963) *Analogue Table*. Photograph courtesy of the Institute of Contemporary Art, Winnipeg

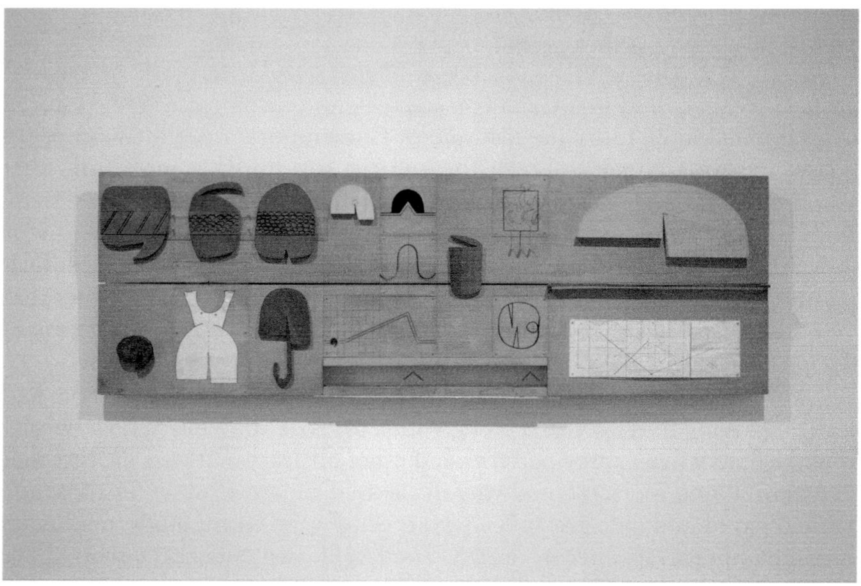

Figure 2.8 Roy Ascott (1963) *Items of Intention*. Wood, Perspex and paper. Photograph courtesy of the Institute of Contemporary Art, Winnipeg

and signals (information in process). *Items of Intention* features playful visual relationships, such as a split curve used to represent a woman's body, and adjacently, the same form as an umbrella. Both enfold, enclose and shelter. There are several drawings in *Items of Intention*, onto squares and strips of Perspex that are attached to the wood, onto paper nailed into place and directly onto the wood itself. These drawings were more notably diagrammatic than the other work of this period; there are grids and vector lines that imply the measurement of a quantity. In the centre, three panels each feature a line with a peak – the top two have a rounded wave to them, the bottom a jagged one. This group of lines relates visually to the recurring motif of a split whole that dominates this analogue – the cleft seeds on the left, the woman's body and umbrella, the split meta-form in dark wood in the centre right, the circle with missing segments on below and to the right of it and the split crescent on the top right with its geometric drawing charting gradients. Out of all the analogues, this has the strongest imprint of Duchamp's *The Large Glass* and Hamilton's later reconstruction of it, particularly in the combination of sharp line drawings with softer abstract form.

The *Analogues* worked on an interplay of analogic relations. They were analogous to Ascott's own material processes and decisions, such as the free lines of the jig-saw cuts. They were, and remain, analogous to alterations made by the viewer, such as the movement of pieces or the sliding of panels. Finally, they are analogous to information flow itself; wave peaks and patterns are *information in movement*, represented in the analogues in peaked and jagged lines. It is this level – the level of information – that was to be the most profound and enduring influence on Ascott's career. Ascott recognized that information was energy, an energy that was transmitted and received, an energy that is dormant in every work of art throughout its existence. From the earliest paintings to the digital image and beyond, the work of art retains its potential to exact change, to accumulate change in reciprocity with interactions over time. Ascott's analogues then, were never complete. To believe in the decisive completeness of a work of art is to believe that it is caught in time, like a fly in amber. Instead, Ascott presented the analogues as works of art caught in *process*, to be made and remade throughout time, in the liberating and boundless exchanges of information that are created by interactivity.

Notes

1 Roy Ascott & Edward Shanken (2003) *Telematic Embrace: Visionary Theories of Art, Technology and Consciousness*. California: University of California Press.
2 Roy Ascott 'Behaviourist Art and the Cybernetic Vision'. In: Ascott & Shanken (2003), 151.
3 Winston Churchill (1949) 'The Wizard War'. *The Second World War Volume II*. New York: Haughton Mifflin Company, 337.

4 Owen Holland and Phil Husbands 'The Origins of British Cybernetics: The Ratio Club'. In: Holland & Husbands (eds) (2008) *The Mechanical Mind in History*. Cambridge: MIT Press, 11.
5 *Ibid.*, 10.
6 See Peter M. Asaro 'From Mechanisms of Adaptation to Intelligence Amplifiers: The Philosophy of W. Ross Ashby'. In: Husbands and Hollands (eds) (2008) *The Mechanical Mind in History*. Cambridge: MIT Press, 151–180, 152.
7 *Ibid.*, 10.
8 *Ibid.*, 10.
9 *Ibid.*, 10.
10 *Ibid.*, 10.
11 Cedric Price (2001) 'Gordon Pask'. *Kybernetes*, Vol. 30. No. 5/6, 819–20.
12 Cedric Price. In: Hans Ulrich Obrist (ed.) (2003) *Re:CP*. Berlin: Birkhäuser, 69.
13 Andrew Pickering (2005) *The Mangle of Practice: Time, Agency and Science*. Chicago: University of Chicago Press, 7.
14 Andrew Pickering. (2010) *The Cybernetic Brain: Sketches of Another Future*. Chicago and London: University of Chicago Press.
15 Andrew Pickering (2002) *Transcript of a paper given in Paris in May 2000*.
16 *Ibid.*, 25.
17 Roy Ascott (1963) *Diagram-Boxes and Analogue Structures*. Catalogue of a solo exhibition at the Molton Gallery, 1963.
18 F.H. George (1959) *Automation, Cybernetics and Society*. Oxford: Pergamon Press.
19 Pethick *Op. Cit.* (2006).
20 Shanken *Op. Cit.* (2003) 10. Ascott quoted from an interview of 1995.
21 Kate Sloan *Interview with Roy Ascott* (September 2012).
22 F.H. George (1962) *The Brain as Computer*. Oxford: Pergamon Press, 1.
23 Tom McGill (2007) 'Design under the Microscope: The Festival Pattern Group 1951: The Council of Industrial Design and the Mechanics of Industrial Liaison'. *The Journal of the Decorative Arts, Society 1850–the Present*, No. 31, 'Influences in Design: An Omnium Gatherum, 92–115.
24 See Rudolf Arnheim 'Gestalt Psychology and Artistic Form' in: Whyte (1951) *Aspects of Form* London: Lund Humphries.
25 Augusto Moreira da Nóbrega and Maria Luiza P. Guimarães Fragoso (2015) 'Field, coherence and connectedness: Models, methodologies and actions for flowing moistmedia art'. *Technoetic Arts: A Journal of Speculative Research*, June 1.
26 Moreira da Nóbrega and Maria Luiza P. Guimarães Fragoso (2015) quoting Rudolf Arnheim (1994) 'The Completeness of Physical and Artistic Form'. *British Journal of Aesthetics*, Vol. 34. No. 2, 109–113, 109.
27 W. Ross Ashby (1960) *Design for a Brain*. London: Chapman and Hall, 36.
28 Kate Sloan *Interview with Roy Ascott* (November 2015).
29 Paul Thirskell (2004) 'From the Green Box to Typo/Typography: Duchamp and Hamilton's Dialogue in Print'. In: *The Tate Research Seminar*, London, UK, 14th July, 2005.: Available online at http://www.tate.org.uk/research/publications/tate-papers/03/from-green-box-to-typo-topography-duchamp-and-hamiltons-dialogue-in-print, accessed February 2016.
30 Jack Burnham (1968) 'Systems Esthetics'. *Art Forum*, September, 1968, 32.
31 Pasmore held an exhibition entitled 'Basic Forms' at the O'Hana Gallery in 1958, showing twelve paintings and constructions. See: Alan Bowness (1960) 'The Paintings and Constructions of Victor Pasmore'. *The Burlington Magazine*, Vol. 102. No. 686, Modern Painting (May), 198–205, 205.
32 See 'Calibrator' chapter.
33 Shanken. Op. Cit. (2003) 13.

34 Kate Sloan *Interview with Roy Ascott* (2015).
35 Ashby (1960).
36 Claude E. Shannon, 'A Mathematical Theory of Communication', *Bell System Technical Journal*, July 1948, 27:3, 379–423.
37 Ashby (1960) 1.
38 See introduction for a fuller discussion of the overlap between psychiatry and cybernetics in Britain. For a summary, see: Phil Husbands and Owen Holland (2008) 'The Ratio Club: A Hub of British Cybernetics'.
39 See B.F. Skinner (1938) *The Behavior of Organisms: An Experimental Analysis*. New York: Appleton-Century; *Science and Human Behaviour* (1953) Macmillan; and *Walden Two* (1948).
40 B.F. Skinner (1938) *The Behaviour of Organisms: An Experimental Analysis*. New York: Appleton-Century-Crofts Inc., 47–8.
41 Cristina Iuli (2013) 'Information, Communication, Systems'. In: Grzegorz Kosc *et al.* (2013) *The Transatlantic Sixties: Europe and the United States in the Counterculture Decade*. Verlag: Transcript. 227.
42 *Ibid.*, Iuli, 228.
43 This was first noted by Charlie Gere, who recognised the deficit of cybernetic work shown retrospectively. See Charlie Gere (2002) *Digital Culture*. London: Reaktion, 373.
44 Jasia Reichardt (1968) 'Cybernetic Serendipity: The Computer and the Arts'. *Studio International*, (special issue) July. London: Studio International.
45 Donna Haraway (1999) 'The Biopolitics of Postmodern Bodies: Constitutions of Self in Immune System Discourse'. In: Janet Price and Margaret Shildrick. (1999) [1989] *Feminist Theory and the Body: A Reader*. Edinburgh: Edinburgh University Press, 203–214, 206.
46 *Ibid.*, 206–7.
47 Yehoshua Bar-Hillel & Rudolph Carnap (1953) 'An Outline of a Theory of Semantic Information'. In Bar-Hillel [1964], 221–74.
48 Hermann Haken and Juval Portugali (2004) *Information Adaptation: The Interplay between Shannon Information and Semantic Information in Cognition*. New York: Springer, 3.
49 Donna Haraway (1989) 206–7.
50 Claude E. Shannon (1948) 379.
51 Gilbert Simondon (1980) [1958] *Du mode d'existence des objets techniques/On the Mode of Existence of Technical Objects*. London, ON: University of Western Ontario.
52 This theme will be explored further in the following chapter. See Andrew Pickering (2010) *The Cybernetic Brain: Sketches of Another Future*. Chicago & London: University of Chicago Press.
53 Shannon (1948).
54 Kate Sloan *Interview with Roy Ascott* (2015).
55 W. Ross Ashby (1957) *Introduction to Cybernetics*, 86.
56 Pickering (2002).
57 Andrew Pickering (2006) 'Beyond Design: Cybernetics, Biological Computers and Hylozoism'. Transcript of a paper given at a conference on the philosophy of technology, Copenhagen, 13–15 October 2005.
58 Ashby (1957) 88.
59 Wiener rebuffed the artist Nicolas Schöffer's request for an interview, replying: 'I'm afraid my interests are more remote from yours than you appear to believe.' See Carlotta Darò for a discussion of their exchange of letters: Carlotta Darò 'Nicolas Schöffer and the Cybernetic City'. *AA Files*, No 69 (2014), 3–11, 5.
60 Gordon Pask 'A Proposed Evolutionary Model' in: Heinz von Foerster and George W. Zopf (eds) (1962) (*Principles of Self-Organization*. Transactions of

the University of Illinois Symposium of Self-Organization, 8 and 9 June, 1961. *International Tracts in Computer Science and Technology and their Application*, 229–253, 229.

61 W Grey Walter's 'tortoises' were light-sensitive automata – they had a 'headlight' and if it reflected off a shining surface such as a mirror, it would attract the tortoises. See Pickering (2010) *The Cybernetic Brain* 50–6 for a full discussion of this automaton.

62 Roy Ascott 'Behaviourist Art and the Cybernetic Vision'. In: Ascott & Shanken (2003), 151.

63 Rosalind Krauss (1979) 'Sculpture in the Expanded Field' *October*, Vol. 8, 30–44, 36.

64 Roy Ascott & Kate Sloan in conversation. Henry Moore Institute, Leeds. March 2017.

65 The recovered analogues were exhibited in 2017 for the first time in the UK at the Henry Moore Institute in Leeds for the exhibition *Form Has Behaviour*.

66 See 'Field' chapter for a fuller discussion of Ascott's use of the jig-saw in 1965–7.

3 Field

Aerial views | Earth mysteries | horizontality

Theories of field and aerial views

This chapter will address the intertwined relationship between field theories and the aerial view, placing Ascott's work in the context of countercultural interests in mapping, charting and 'reading' the landscape, as well as exploring his subsequent interest in tables and the horizontal plane of human communication. As a young man growing up in the vicinity of Glastonbury, Ascott developed an awareness of the beginnings of the 'Earth Mysteries' movement, which brought together ancient monuments and science fictional speculation in searching for pattern and meaning in the landscape. Taken in combination with Ascott's vital years in Fighter Control, these more spiritual approaches to the aerial view fuse with the communication fields produced by contemporaneous technologies in his early work. The aerial photograph, the map and field theories of consciousness together express the multiple, and often surprising, interactions between technological development and cosmic speculation that informed the British countercultural movement.

Over the course of the twentieth century, the term *field* became cross-disciplinary, used to denote open spaces of exchange. Its original geographical meaning of a large tract of open and flat land translated to a conceptual zone of interaction, involving information, ideas, languages, cultures or behaviours. In mathematics, in the sciences and in engineering, *field* was employed as a phrase to express extended boundaries of practice against the growth of increasingly systematized technologies and methodologies.[1] Donna Haraway has argued that 'The single most important organizing principle developed in this century to treat pattern was that of field', an observation that connects 'field' with observation and the identification of pattern.[2] In this sense, 'field' was an abstraction of systems thinking, a way to express the external factors that contribute to the function of any individual organism, mechanism or person. Hayles describes the 'field view of reality' as one in contrast to the 'atomistic Newtonian idea of reality', a view that 'pictures objects, events, and observer as belonging inextricably to the same field'.[3] In this world-view, each of these elements is effected by any change in the other.

Haraway's outline of a field theory for the sciences is equally applicable to the arts, to sociology and to psychology; each of these disciplines evolved field theories of interaction in the twentieth century. It is no coincidence that this interconnected, spatial, theoretical analogy became prolific during the century that allowed us, for the first time, to view the world from above; first from the aeroplane and then from the spaceship. These zoomed-out world-views created a new geography, both local and universal; the aerial view allowed distance, allowed us to develop a connected vision of being. Field theory is essentially pattern-seeking; it takes in the external environment around an object of study, whether that object is a machine, an artwork, an organism, an individual, a star or a city. As the multiple theories of vision that had emerged following the war filtered into the visual arts, so did the concept of a field of vision. James J. Gibson, György Kepes and László Moholy-Nagy had each written books that had been widely read by artists across the world, each book in its own way contributing to a field theory of vision.[4]

Roy Ascott was amongst the many artists of the 1960s who began to use the term in relation to painting and sculpture. In 1961, Ascott stood with a critic before one of his analogues which was titled *Field*, on exhibition at the Molton gallery. It had background of green baize – the critic asked Ascott whether that green indicated the 'field' in question, reading it entirely literally as a response to a rural scene.[5] This little exchange highlights the complications of an evolving vocabulary of visual art, as well as the fact that 'field' was yet to be fully integrated into this vocabulary. It also demonstrates that by 1961, Ascott was using the term field in relation to his analogues to express an extended space of interaction. This was to be a critical interdisciplinary concept. By the 1990s, Pierre Bourdieu had formalized a theory of *field* in relation to *habitus*, his critical theory of human agency and power.[6] Principally a sociological study of the formation of habit and identity, Bourdieu also wrote on the production of art within the field, writing: 'The meaning of a work (artistic, literary, philosophical etc.) changes automatically with each change in the field within which it is situated for the spectator or reader.'[7] Bourdieu's broad conception of field behaviours in creative production had much in common with Ascott's distinctive philosophy of visual arts practice; both were broadly rooted in the concept of feedback which by then had become absorbed into interdisciplinary discourse, detached from its cybernetic origins. Ascott had articulated a field theory for the visual arts in 1980, which would 'replace the formalist modernist aesthetic', and which focused on behaviour rather than form. He wrote that this theory:

> looks at a system in which the artwork is a matrix between two sets of behaviours (the artists and the observer) providing for a field of psychic interplay which can be generative of multiple meanings, where the final responsibility for meaning lies with the viewer.[8]

Ascott intended his field theory to be a theoretical model for postmodernist visual art practices, in which the artwork was a start point rather than end point, and in which, with an echo of Barthes, the viewer ultimately held the power of attributing meaning.[9] In essence then, for Ascott the term 'field' expressed the complex relationships between frameworks of production and dissemination in the visual arts, which both inform meaning and are altered by meaning, in the same way that Bourdieu's *habitus* forms, and is formed by, human behaviour.

The view from above: Ascott, Maltwood and the aerial photograph

Ascott spent his childhood in Bath, and from a young age he was aware of the alternative culture that existed in and around nearby Glastonbury, a place that held simultaneous mythological, archaeological and creative status throughout the twentieth century and since. While the heyday of Glastonbury's alternative culture was the 1960s and 1970s, from the inter-war years the area was a draw for artists. One such artist was Katherine Maltwood. She was a British artist whose interests included the occult, astrology and spiritualism. She trained in the UK at the Slade at the end of the nineteenth century and worked in the UK until she moved to Canada with her husband in 1938, where she remained for the rest of her life.[10] Given the dates of her training, Maltwood's interest in spiritualism was very much of its time, particularly with regards to its place within the Arts and Crafts movement as well as the Celtic Revival in Scotland.[11] Maltwood was, however, rather more unusual in that her interest in spiritualism, the occult and astronomy was to endure for her whole career.

Just after World War I, while in pursuit of the Arthurian Grail legends, Maltwood had created a series of images that are now firmly built into the fabric of Glastonbury's longstanding alternative culture. King Arthur and other characters from the legends were depicted numerous times by the artists of the Arts and Crafts movement, including Edward Burne-Jones and William Morris. As a result of the revival of interest in the Arthurian legends, Maltwood had been commissioned in 1925 to draw a map that would illustrate the tales of King Arthur's knights and the Holy Grail, so she began to amass ordinance survey maps and also the very new resource that was aerial photographs. Exploring the landscape of Ascott's childhood, she traced out routes, paths and symbols over the land of the South West of England. Maltwood cross-referenced her photographs with '6 inches to 1 mile ordinance survey maps', tracking roads and buildings as well as topographical features of the landscape.[12] It was while gazing down at these paper aerial landscapes that Maltwood made her discovery – she saw before her a circular map of the Zodiac surrounding Glastonbury. She claimed that 'Looking down upon them from the air, with the aid of these maps, it can be seen that they delineate enormous effigies resembling Zodiacal creatures arranged in a circle' (Figure 3.1).[13] Ascott himself firmly recalls that Maltwood had, in fact, taken to the skies herself, strapped in a small plane, in order to make this discovery, an anecdote that at the very least confirms the romantic place Maltwood held within the folklore of British counterculture.[14]

Whether the process of envisioning the zodiac took place during physical or mental flight, Katharine Maltwood created an aerial view of Glastonbury and surrounds that would make a considerable contribution to its identity as a spiritual place that was built upon a mythical – and magical – landscape. Maltwood prepared drawings, creating silhouettes of the symbols of the Zodiac, each suggested by hills, rivers, roads and gradient lines. The resulting images operate around a tentative gestalt; each creature of the Zodiac is filled out in black, its outline traced from new and old features of the landscape that could be seen on the map or from the sky. Together they form a wheel with Glastonbury in the centre. In black and white, the circle of the Glastonbury Zodiac looks compellingly similar to the hazy black and white target screens that would grace fighter jets by the end of World War II.

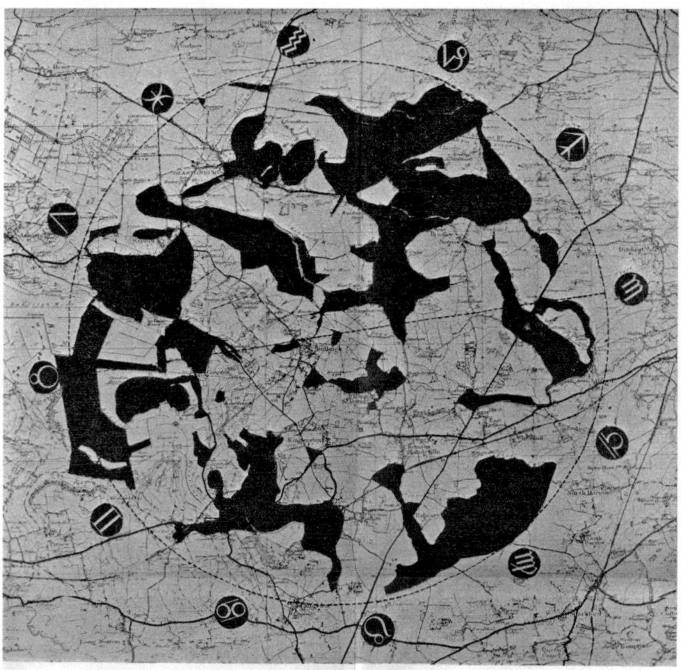

KING ARTHUR'S ROUND TABLE OF THE ZODIAC
Somerset, England

Figure 3.1 Katharine Maltwood (1929) *The Glastonbury Zodiac*. Illustration from Maltwood (1929) *Glastonbury's Temple of the Stars* Cambridge: James Clarke and Co.

Ascott had become fascinated by this strange collision of spiritualism and the aerial view that was famous in his home region. Later, while teaching at the University of Wolverhampton, his programme of events reflected his countercultural preoccupations with mystic landscapes, consciousness and communication. At this time, he had a developing interest in the paranormal or, more accurately, the parapsychological. Amongst other speakers he invited in this era was Queenie Nixon, who Ascott remembers as 'the last of the transformation mediums in Britain'.[15] She performed a transfiguration demonstration for students, at which, Ascott recalls, her own face appeared to melt, stretch and distort like 'the clock in Dali's Persistence of Memory', as she was possessed by a series of visiting spirits. A female student recorded this in photographs, having tried unsuccessfully to film the event. Ascott saw the resulting photographs, which he remembers as having showed faces floating all around Nixon's own head, some distorted like melting dolls.[16] Ascott's interest in Nixon connected to his broader interest in the mysterious field of consciousness and communication.

For Ascott, Maltwood's maps channelled the pure magic of the aerial landscape and the radically different view of the world it allowed. He was not cynical about Maltwood's enterprise, seeing it as an example of the ancient and curious secrets that were gradually revealed to the world through aerial photography; for example, the new visibility of archaeological features that while easily missed on level ground, became clearly evident from above. Ascott was also interested in the archaeological site at Avebury, which contains the largest stone circle in Europe, as well as the nearby Silbury Hill, a prehistoric manmade chalk mound (Figure 3.2).[17] Ascott was attracted by the mystery around the function of such sites, recalling that 'Nobody has any idea why it's there. No tombs, the structure is brilliant. It won't erode, it's been there for thousands and thousands of years and its part of an enormous complex [...].'[18] Avebury's scale and design also prompted Ascott to believe that the monument was designed to be viewed from above, a thought that was also integral to the more mystical interpretations of this monument over time, from the Druidic to the Arthurian.

During Ascott's tenure at Wolverhampton, the author John Michell gave a talk on ancient pathways, based on his 1969 book *View Over Atlantis*.[19] This book had drawn from Alfred Watkin's 1925 book *The Old Straight*

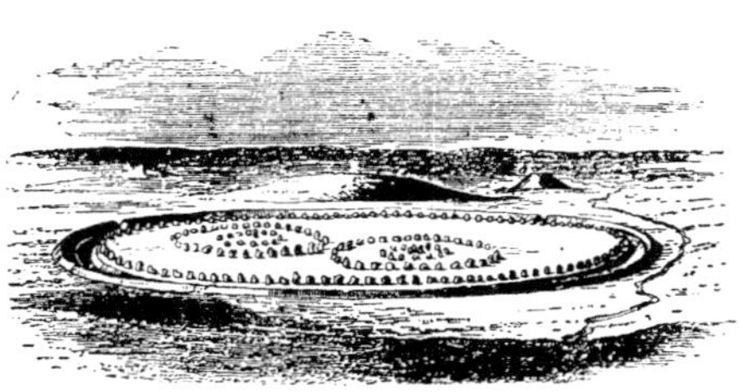

Figure 3.2 John Martin *Avebury*. Published in *Nordisk Familjebok*, nineteenth century. Based on an illustration by John Britton

Track: Its Mounds, Beacons, Moats, Sites and Mark Stones, which had coined the term 'ley lines' and outlined a detailed theory.[20] Watkin had viewed the landscape from a Roman camp, perceiving that there were trails between the ancient features that were scattered across the landscape, a network of what he would term ley lines.[21] Like Maltwood, Watkin and Michell believed in an ancient and mystical geography that could be perceived from above, still present in the changing landscape of twentieth-century Britain. In *The Flying Saucer Vision*, Michell made the following extraordinary claims, worth quoting in full:

> It is impossible to study the evidence of ancient metrology without forming the conclusion that the various units of weight and measure, many of which survive today, were originally part of a remarkably evolved system which could only have been conceived by some advanced scientific civilization in the remote past [...] The laws of harmony and proportion were considered to provide the key to the secrets of creation, and those laws were followed in the construction of a hierarchical society, in the arts and sciences and in every other aspect of life. These were the divine laws which, according to Plato guided the people of Atlantis, and when they were neglected decadence and destruction were the inevitable consequence.[22]

Michell described the beauty and balance of ancient roads, monuments and buildings as a kind of code, which he believed could only have been conceived in a previous scientific age, lost in time. The theory of a previous scientific age recurred in counterculture, in part a measure of the alienation and cultural change provoked by rapid technological progress. In Michell's *The View Over Atlantis*, he began with Glastonbury and the ley lines connecting the many ancient features of the surrounding landscape, taking the idea of sophisticated past design further by assigning meaning to the structure and proportion of the selected features. In a collision of mysticism and science fantasy that was all too common at the time, he used this interpretation of the landscape to build an argument for the existence of UFOs, as well as the loss of highly potent, ancient knowledge. The book was in fact part of a trilogy, which also included the titles *City of Revelation* and *The Flying Saucer Vision*.[23] This trilogy inspired the beginnings of the Glastonbury Fayre, which would later become the Glastonbury Festival.[24] Patrick F. Sheeran has described the rise of the Earth Mysteries movement in British counterculture:

> Two central ideas, that sacred places form a pattern in the landscape and that they are associated with some form of energy, became the basis for a whole range of studies, fantasies and conjectures. The shelves of alternative book shops all over Britain carried a new section, somewhere alongside Ecology and Esoterism, called Earth Mysteries.[25]

The discussion of hidden codes in the landscape and forgotten contact with other planets and alien beings is very much of its time – a collision of the ancient and the very modern set against the cultural backdrop of the space race. Given the progressive quickening of twentieth-century technological change, the slow landscape that followed ancient paths, monuments and natural features held a quite obvious attraction for artists, writers and musicians of the era. From industrialization to the technological age, modernity was a brief explosion in a far-reaching history. In this context, the popularity of the notion of other, forgotten technological ages lost in time, that preceded our own, is understandable. It was directly related to the speed of progress of the twentieth century, which in turn provoked a sense of time itself quickening, rushing forwards. Sheeran also noted that it '[…] is surely suggestive that the leading role in the development of geomancy this century was taken in Germany in the 30s and Britain in the 60s – by two nations with imperial aspirations, one briefly resurgent the other in decline'.[26] This kind of geomantic mapping then, has implications of the revival of ancient powers in the face of imperial decline.

It should also be noted that the mystical landscapes that occupied Ascott and his contemporaries in the 1960s and 1970s share the common factor of the aerial viewpoint – given that Ascott had worked in Fighter Control, this creative and personal interest in landscape as spiritual maps bears an interesting connection to his military service. From the map of the Ground Control bunker to the aerial views collected by radar and by pilots, the remote view from above evolved for Ascott from both technological and spiritual points of origin. With Avebury in particular, the cult belief that it might have been a message directed to the skies, like crop circles, was part of Ascott's broader and enduring interest in prehistoric culture.

Glastonbury, then, accrued layered myths and meaning in this period, augmented by the alternative mappings, visual and textual, created by Maltwood, Watkins and Michell. Andy Worthington has described how the influence of Watkins and Michell in particular, encouraged groups to visit sites and routes in the area, tracing these Earth mysteries:

> […] the UFOlogist Tony Wedd, whose speculations about UFO activity in the countryside of southern England led a small group of aficionados to rediscover the works of Alfred Watkins, establishing the Ley-Hunters Society and turning up at Avebury on a field trip in 1962. Undergoing a transformation that mirrored the counter-culture in general, the ley-hunters opened a path for increasingly cosmic field trips, resulting, in 1969, in the publication of John Michell's *The View over Atlantis*. This, the founding document of a movement that became known as 'earth mysteries', revived Avebury, Glastonbury and Stonehenge as centres of mystical power, and introduced themes of geomancy, earth energies and sacred geometry which began to exert a powerful influence on those interested in both archaeology and spirituality.[27]

It can be little surprise then that some relationships developed between alternative interests such as astrology and parapsychology and the landscape – the engineered possibility of flying through space echoes the interests in disembodiment that was an important facet of all strains of spiritualism. It had not yet been a century since the human race had taken to the skies; the world as viewed from above had only been possible since the invention of the aeroplane. In the interwar years, aerial photographs became available, offering a new view of the world below. The curious patchwork of fields, roads, the nerve centres of cities, the strange visual response to seeing vast mountain ranges reduced to wrinkles on the surface of the Earth.

The problem – or possibility – of scale that this created was reflected in a number of the Independent Group exhibitions, including *Parallel of Life and Art* at the ICA in 1953, in which aerial photographs jostled with images of organisms, archaeological finds and works of contemporary art, all homogenized through the medium of black and white photography (Figure 3.3). In an internal memo discussing the exhibition, Nigel Henderson had commented: 'Technical inventions such as the photographic enlarger, aerial photography, and the high-speed flash have given us new tools with which to expand our field of vision beyond the limits imposed on previous generations.'[28] Zooming in and out, the organism could resemble a landscape, the landscape a fine tracery of nerves.

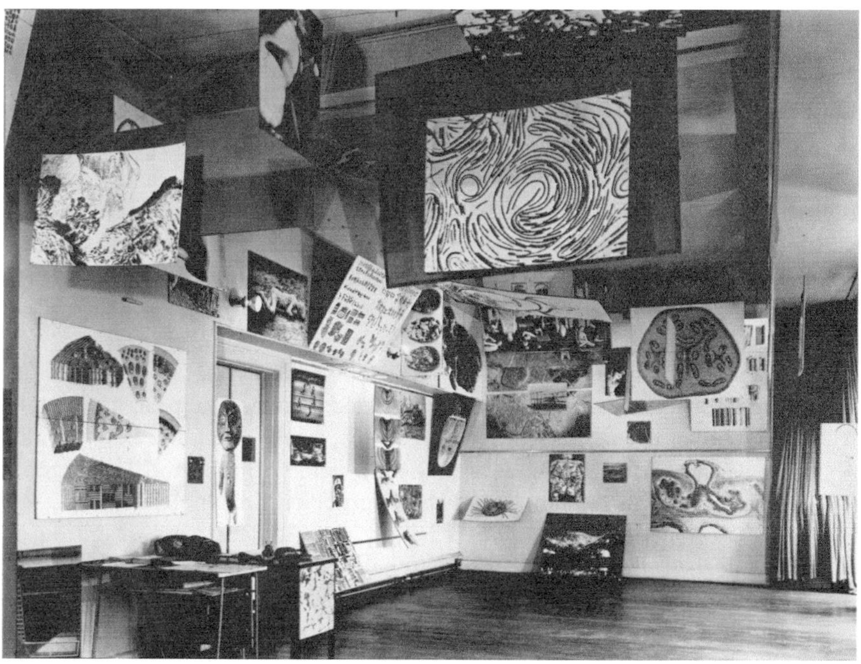

Figure 3.3 Eduardo Paolozzi and Nigel Henderson (1953) *Parallel of Life and Art*, London: ICA. Photograph: Tate Archive

The same cross-fertilization of imagery was evident in both Richard Hamilton's *Growth and Form* exhibition at the ICA during the Festival of Britain in 1951 and György Kepes' compellingly similar exhibition *The New Landscape in Art and Science*, which took place the same year in the United States.[29] Central to these innovatory crossovers between zoomed-in and zoomed-out imagery were the technologies that allowed these new visions of the microcosmic and macrocosmic world to develop, from photomicrograms to the aerial photograph. World War II had provoked a surge of activity in mapping the world from above, and in as much secrecy as possible. Accurate aerial maps were a resource that transformed aerial warfare, providing pilots with accurate views of targets in advance of missions. The missions to take the photographs were dangerous in themselves, a form of vulnerable espionage. Back in the UK, film was developed in darkrooms in secret locations, including the Hughendon Manor's Ice House.

At Hughendon Manor, thousands of surveillance photos taken by pilots over Germany were translated into maps. The prints were overlaid with concentric circles to indicate the targets, each circle a mile apart (Figure 3.4). These maps were geographies of intention, charged with the remote power of the aerial view. The aerial view of the world has become intimately familiar over time, but during and immediately after the war, the world as seen from above still maintained a curious and unfamiliar place in visual culture. The artists Harold Cohen and Bernard Cohen were both involved in Groundcourse, and at the start of the 1960s their work exhibited a topographical aesthetic that was a common and compelling strain of late British abstraction. Bernard Cohen's *Floris* (Figure 3.5) was apparently inspired by the elaborate icing designs of the bakery of the same name but it also resembles a topographical drawing of a mountain range, with the concentration of lines spreading outwards from the mass in the bottom left of the image. The red lines also resemble electrical pulses or readings, along with the irregular white dots which float beneath the surface. There is a cluster of more regular white dots at the centre of the mass, like nodal points on an electrical circuit.

The painting was evidently built up in layers, with shadow forms in grey and a block of darker blue under the surface map of red. This layering conversely creates a sense of flatness rather than depth, blocking in the visual field and making it shallow. While this painting was apparently worked on the vertical and displayed on the vertical, it retains a distinctly cartographical look. Similarly, two lithographs from a series by Harold Cohen called *Close Up* resemble landmasses seen from the air. *Close Up I* looks like islands, depicted in areas of stark red, blue and black (Tate Collection). The surface of the print has a secondary pattern of shadows, much like cloud shadows upon the surface of land or sea viewed from above. This print hovers somewhere between landscape and microscopy; it could also be a pattern of blood cells, for example. In this way it overlaps landscape, biology and abstraction in a manner that recalls *On Growth and Form*, *Parallels of Life and Art* as well as Kepes' *New Landscapes of Art and Science*. Much of

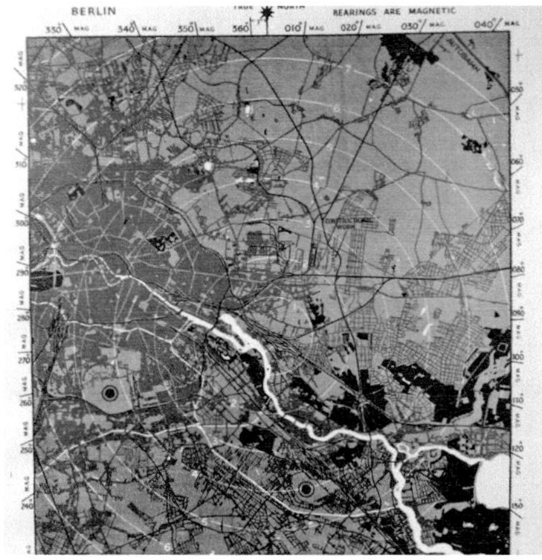

Figure 3.4 Map produced in the Hughenden Manor Ice House. Photograph © Claudia Droel

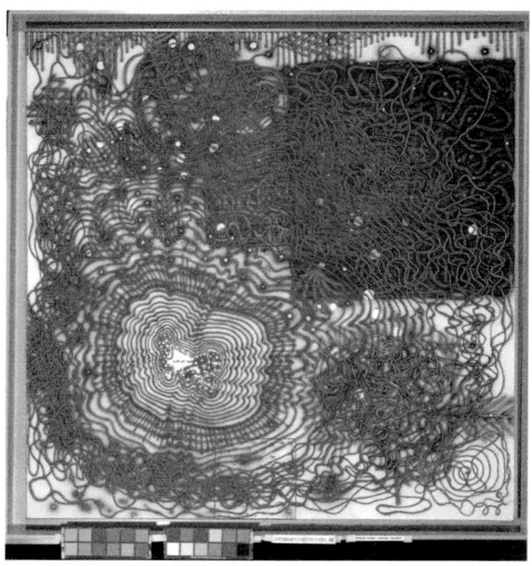

Figure 3.5 Bernard Cohen (1964) *Floris* oil paint and tempera on canvas, 1830 × 1834 mm, Tate Collection

Harold Cohen's work in the 1960s employed the same device of confused scale, including *Vigil Completed* (1966) and *First Folio B* (1965).

In a second print five years later (*Close Up V*) Cohen uses a darker palette, suggestive of a night-time scene of the same subject. The title of the series implies a direct engagement with the issue of scale, since the prints hover between the zoomed in world of microbiology and the zoomed out and distant world viewed from an aeroplane. In this print, two peninsula forms extend upwards from below, and a third extends downwards from the top right of the image. The top left quarter is submersed in shadow, concealing a grid of regular dots beneath it. Grids, dots and other divisions and markers were part of the visual lexicon of World War II landscape imagery, in which topography was charted, measured and targeted. The aerial photograph recalls the detachment of the pilot for those on the ground, the strangeness of the new age in which we could rise above the landscape that had surrounded us for so long. This was connected to scale – the vastness of cities is reduced, in aerial format, to complex and minute diagrams that double vastness back, that make cities resemble strange and organic microcosms, thereby reducing the grandest of human ambitions to frailty. The surveillance map reflected a new geography of warfare, in which precise addresses could be pinpointed with ease. It was a step on the journey to our own age of surveillance. We no longer live in a world where we can feel enclosed by place, by the physical miles that surround us. Technology has intervened, rendering distance as plastic, easily overcome.

Cosmic epiphanies: the Earth from the stars and Ascott in the US

In 1971, the astronaut Edgar Mitchell was floating high above the Earth aboard Apollo 14, on its return journey from the moon. As he gazed down at the planet floating in space, he experienced an epiphany, a moment of clarity in which he sensed a 'universal connectedness', the deep and mysterious potential of the field of consciousness. When he returned to Earth, he wanted to make a profound change to his life's direction, one that would allow him to explore this epiphany in a productive way. This was what provoked him to found the Institute of Noetic Sciences in California, an organization 'dedicated to supporting individual and collective transformation through consciousness research, transformative learning, and engaging a global community in the realization of our human potential'.[30] Mitchell had determined an ambition 'To broaden the knowledge of the nature and potentials of mind and consciousness and to apply that knowledge to the enhancement of human well-being and the quality of life on the planet'.[31] 'Noetic' has its roots in the Greek word *noetikos*, which in turn relates to the mind and to knowing, applied in this context as intuitive knowing. In essence, the organization was founded on Mitchell's profound new belief in the interconnectedness of things, from the human to the celestial. He '[…] realized that the story of ourselves as told by science—our cosmology, our religion—was incomplete and likely flawed', arguing that '[…] the Newtonian idea of separate, independent, discreet things in the universe wasn't a fully accurate description. What was needed was a new story of who we are and what we are capable of becoming.'[32]

Mitchell's venture was in its early years when Ascott took up a post as vice-principal at the San Francisco Art Institute, before being swiftly promoted to dean. This was how it came to be that Ascott received a phone call inviting him to contribute to proceedings at the Institute of Noetic

Sciences. While visiting the institute, Ascott witnessed, amongst other curiosities, the blind Brazilian artist and psychic Luiz Gasparetto, who could replicate four different works of art by four different artists simultaneously, using both hands and both feet.[33] Ascott himself held the paper in place as Gasparetto drew, apparently channelling the souls of the deceased artists. Ascott recalled the curiosity of watching famous drawings emerging simultaneously; a Van Gogh would evolve alongside a Cezanne, a Monet, a Renoir [...].[34] There is a film of the young psychic from another demonstration in 1978, broadcast on the BBC in a dedicated episode of the television programme *Nationwide*. It was entitled 'Renoir: Is that You?' and it shows the artist drawing frantically, a hand cupped over his eyes, as a portrait of a young woman in the style of Renoir emerges in charcoal.[35] This particular strain of psychical performance, in which the medium embodied the spirit of deceased artists, was popular in the 1960s and 1970s. Gasparetto's work reflected the Institute's pursuit of intuitive creativity beyond the boundaries of perception, even beyond the senses. In the time of Ascott's involvement, they were busy with a project assessing and measuring human levitation, an interest that had a following within both British and American counterculture, a fact that has interesting resonances in the context of the aerial developments discussed in this chapter. Mitchell's Institute strayed regularly onto the territory of parapsychology in the search for an expanded field of consciousness. Ascott's field theory connected with these ideas; he suggested that art should be '[...] understood as a field of psychic probability, highly entropic, in which the viewer is actively involved, not in an act of closure in the sense of completing a discrete message from the artist [...]'.[36]

It is ironic that Mitchell's venture into intuition beyond current scientific knowledge came about from the ultimate triumph of science and technology in the form of space travel, one of the defining moments of the twentieth century. As Pamela Lee has noted, the televised images of the crew of Apollo 11 stepping onto the surface of the moon were some of the most profound images 'in the history of technology, offering a view of the 1960s buoyed by the Enlightenment platforms of reason and progress'.[37] Lee describes how this triumphal vision of man overcoming time and space filtered into public consciousness in blurred and grainy form through the remote, flickering presence of television sets in domestic settings.[38] This iconic moment in the history of the twentieth century was enabled by technologies that allowed us to communicate from the moon to the Earth, as well as directly into private domestic worlds. The uncanniness of this transmission rests on that collision between fathomless space and the confined limits of domestic life.

There are only a few hundred people who have looked back on the Earth from space and who have experienced the view from above that was so transformative for Mitchell. In essence, Mitchell rejected the same scientific principles that had given him the skills and education to join the crew of Apollo 14, while still suspended above the Earth. Seized by a sense that there was an energy connecting everything, he strayed from the rationalist view that consciousness was born of matter and died with it, believing instead in a sweeping and harmoniously systematized view of all life, from the organismal to the intergalactic. In a recent interview, Mitchell reminisced about travelling back towards the Earth in a spacecraft that was rotating to maintain 'the thermal balance of the sun', so that every two minutes the astronauts saw '[...] the Earth, the Moon and the Sun as they passed by the window'.[39] The unimaginable brightness of the stars triggered a profound joy, in which Mitchell '[...] realized that the molecules of my body and the molecules of the spacecraft had been manufactured in an ancient generation of stars. It wasn't just intellectual knowledge – it was a subjective visceral experience accompanied by ecstasy – a transformational experience.'[40]

In the many curious overlaps between science and counterculture in the 1960s and 1970s, Mitchell's journey into studies in the field of consciousness was not an oddity.[41] Further, the interconnected study of matter, life and consciousness at the heart of the field of technology studies, held in close communion with the dichotomy of hardware, software and information.[42] The enclosing simplicity gained by framing life within the confines of Newton's second law of thermodynamics, with a return to entropy at the end of life, does not work for any systematized view of living.[43] Writing about the French philosopher Gilbert Simondon's theories of the individuation of life, Elizabeth Grosz has observed that 'Life is not a special kind of substance, a vital force that must be definitively distinguished from matter. Rather, for Simondon as for Bergson, life is a deviation of matter, one of the forms that matter generates.'[44] Grosz goes on to discuss the problems of definition for individuation of life:

> Life is not a difference in kind from matter (as Bergson suggests) but a difference in degree; the living never attain the cohesion and unity of the material individual that 'crystallizes' all it needs of its pre-individual forces at once. There is no moment of attaining an individual, self-identical or stable state which dramatically transforms pre-individual forces, the disparities in potential energy between incommensurable and non-communicating forces, into fixed individuals, as occurs chemically in quantum-type leaps of molecular reorganization.[45]

These discourses revolve around Simondon's lifelong interest in both natural and cultural processes, and particularly in challenging the 'classical oppositions of the vital to the mechanical'.[46] Many of these debates have informed Ascott's enduring interest in consciousness studies, with the questions around life, matter and consciousness recurring in the journal *Technoetic Arts: A Journal of Speculative Research*.[47] While Mitchell's *Institute for the Noetic Sciences* was founded upon, and continues upon, definitively experimental and alternative modes of parapsychological investigation, it is also evident that scientists, artists and others working in and around ideas of expanded consciousness in the 1960s and 1970s were intuitively exploring something sensed rather than proven – that existing rational expressions of life, death and consciousness insufficiently expressed the possibilities presented by recent technological developments. In a letter received by Ascott in 1971, his friend and former colleague Victor Pasmore commented that '[...] whether or not we are approaching a metamorphosis, it does look as if the visual arts will follow science in its extension from the concrete and physical to the mental and psychic', a more cautious expression of the role of science in the development of conceptual art.[48]

It is evident then, that there was a tangled and curious relationship between technology, aerial vision and theories of consciousness, perhaps a simple reflection of the elegiac beauty and mystery of straying, briefly, outside our world, hovering above. The astronaut then, might experience insights generated by exclusion in a similar vein of romantic isolation as the hermit, the castaway or the solitary adventurer. So much post-war art engages with the extended reach, strength and speed of the body in relation to modes of transport – Hamilton's *Man, Machine and Motion* exhibition was a good exemplar of this take on the cyborg, reflective of the Independent Group's techno-modernist engagement with cybernetics in the 1950s.[49] Many cybernetic artists, including Ascott, pursued something else – the extended reach of the mind and the abstraction of communication into an information field. This distinction rejected the suggestive power of a cyborg-human, extended to god-like status by engineering. Instead, the human in a communication network is a node, picking up signals in a field of consciousness which had a vastness great enough to overwhelm, rather than extend, the individual – not unlike an astronaut, gazing down at the remote Earth .

Horizontal games

In 1965–7, Ascott began to experiment with what he described as 'parameters of chance' (Figure 3.6).[50] He placed large pieces of block board on the horizontal, and drew swift, scribbled patterns over the surface, looping and intersecting. He would then throw coins at the drawing, before using a jigsaw to cut along the lines between the coins, allowing a shape to emerge. Sometimes, the block board would end up in pieces, and other times, an intricate shape would emerge.[51] *Random Map I* and *Random Map II* (Figure 3.7), both created in 1967, were created using this process.

This blend of intuitive process and decision-making resulted in wall-mounted sculptural works that have the same endlessness as the aerial map. The cuts create a topography, incisions creating peninsulas and inlets with meandering grace. Writing in 1967, Ascott played upon the visual ambiguity these works created, using pale, uneven colour to further blur their boundaries:

> The use of washed colour reinforced the linear quality of the cuts (where opaque colour would have made them the edges of the forms). These

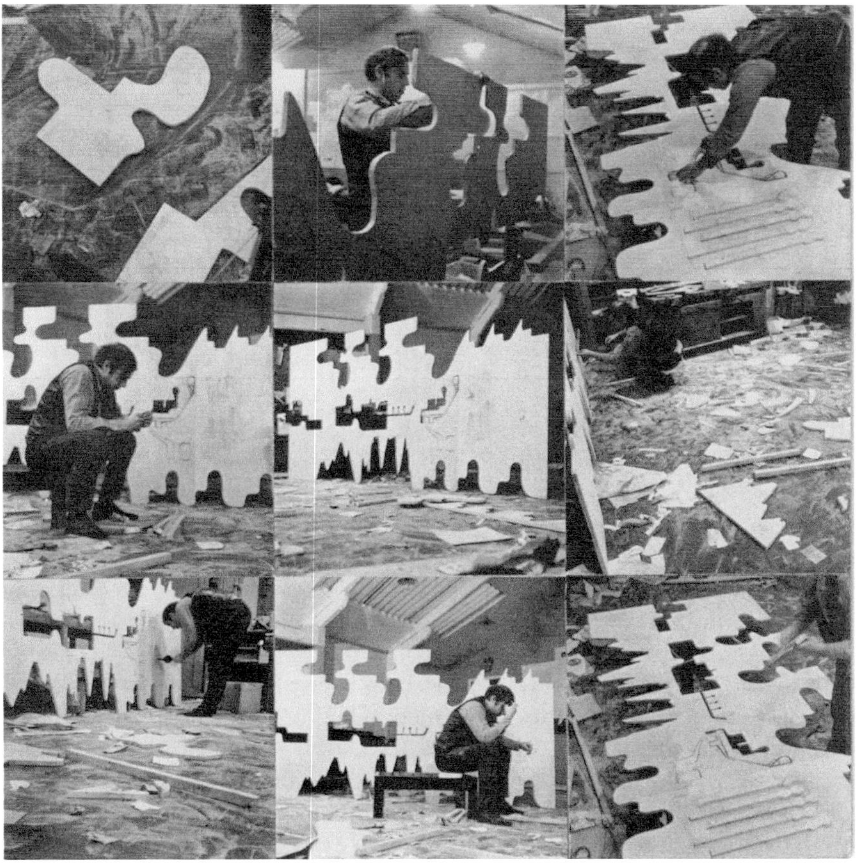

Figure 3.6 Roy Ascott's Studio (1965) Photograph courtesy of the artist

> were essentially *linear* analogues discussing and creating ideas of *modulation* in terms of wave forms, fretted profiles, vector diagrams, and oscillating edge-rhythms.[52]

These ambiguous maps were created through the combination of chance and process, and as such they create a perfect tension of uncertainty and definition. Ascott describes the contour lines as 'vector diagrams', also mentioning the 'wave forms' which were another preoccupation of his at the time, inspired by the wave readings of curtain radar that he had witnessed during his service in Fighter Control.[53] Shanken has noted that in several works of the 1960s and 1970s, Ascott 'visually suggests equivalences between *I Ching* hexagrams, binary notation of digital computers, scatterplots of quantum probability, wave forms of information transmissions, and biomorphic shapes'.[54] Certainly, the throwing of the coins directly recalled the

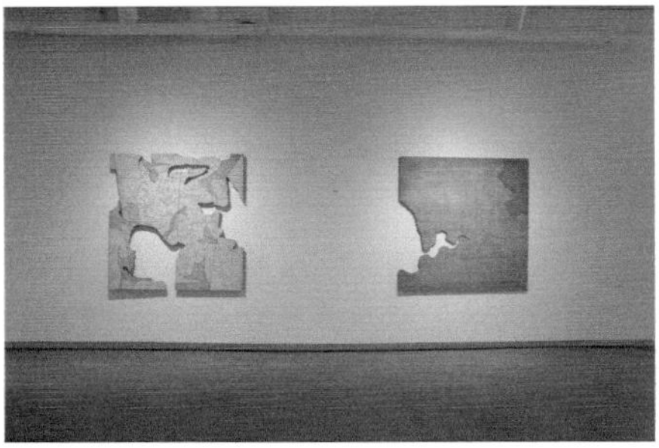

Figure 3.7 Roy Ascott *Random Map I* and *Random Map II*. Photograph courtesy of the Institute of Contemporary Arts, Winnipeg

I Ching, a form arising from the patterns of the coins. This was a mixture of chance and control, a defined gesture and surface but an unplanned result each time, that recalls Ascott's early interest in Pollock.[55] Ascott recalls he was becoming increasingly convinced that art had come 'off the wall', an idea that he would soon take further with his table-top pieces.[56] Shanken has written of this transition to the horizontal as expressing 'the ongoing reconceptualization of painting from a "window on the world" to a cosmological map of physical and metaphysical forces'.[57]

In *Cloud Template* (Figure 3.8), Ascott rotated the plane of the block board on the wall, so the corners formed a diamond, with the top and bottom points flattened. The cut-out shapes are indeed clustered like cloud formations, or like an archipelago viewed from above. The cloud analogy played upon the chance arrangement of form, the liquid way the sky changes in relation to environmental factors. *Cloud Template* trapped and froze a process; as Ascott would argue, it was an analogue of a process, or what Shanken has described as the development of 'change patterns'.[58]

Ascott started working with these ideas for a 1965 solo exhibition at the Hamilton Galleries in London, writing that he was working with new ideas including '[...] the synthesis of low definition content and an *implicitly* flexible container, or more precisely, a formal condition in which the *container was the content*'.[59] This idea holds echoes of the perfect alignment of gesture and mark that Ascott had so admired in Pollock's work, but it also reflects the definition of analogues as machines which accurately and precisely reflect the physical quantity which they measure. Ascott also saw the art object as a container for the viewer's ideas, particularly those works with interactive elements. He describes this:

> *Fourier*, a blue square with wave form edges, turned first inwards diagonally and then outwards towards the spectator. The large blank areas of stained board served as *receptacles* for spectator ideas, with the general context shaped by the linear configurations that contained them.[60]

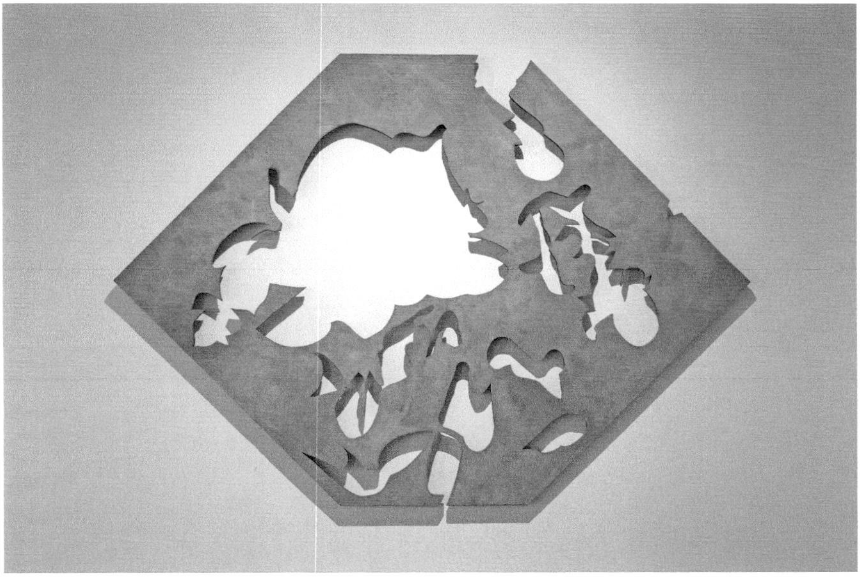

Figure 3.8 Roy Ascott (1967) *Cloud Template*. Photograph © Roy Ascott

Through a formal process of aesthetic decision-making involving positioning, unfolding or sliding, the viewer-participant imposed a new order on the work, which became a 'receptacle' for these meanings. Ascott was articulating the responsiveness of the work of art – the fact art in itself had 'behaviour', explaining that 'present experiments with behavioural structures include the use of the spectator's *approach* and *proximity* to the work to cue changes and movement within the artefact'.[61]

After the free gestures of throwing, Ascott used the jig-saw as a drawing tool. He noted in 1967 that:

> The works are also records of certain random actions and motions with a powered fretsaw, where sawing becomes a way of demarcating one's 'arena'. This topological tracing of paths with a saw across a field at times generates new goals, which come out of the process.[62]

The saw formed 'pathways' across the topography, indicating how clearly Ascott identified this group of work with mapping and with the powerful idea of ley lines central to the Earth Mysteries movement. Ascott used the jig-saw to draw paths that would highlight the topographical 'features' of the coin locations. In *Parameter IV* (Figure 3.9), the black outline recalls radar maps in which landmasses and oceans appear in stark relief. There is also an echo of the gestalt exercises that were part of the Basic Design movement, in which a careful balance of black and white was explored in

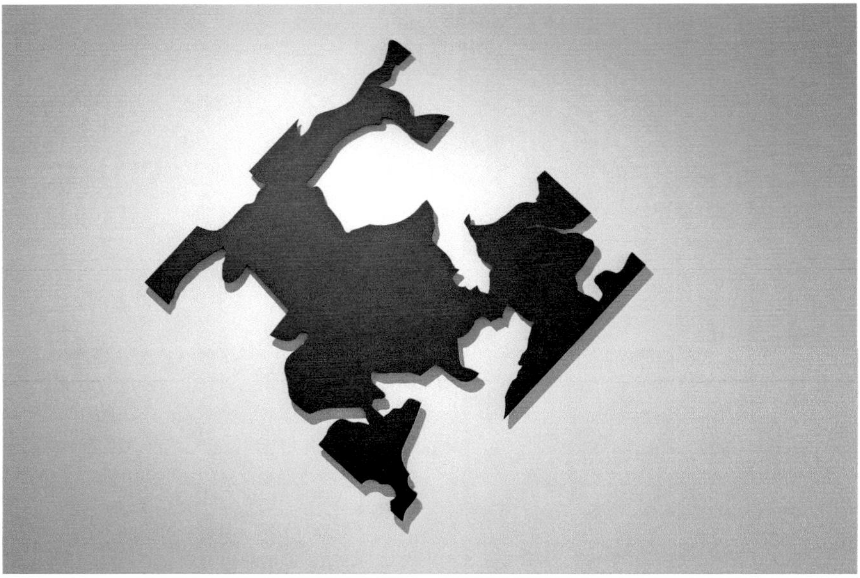

Figure 3.9 Roy Ascott (1967) *Parameter IV*. Photograph © Roy Ascott

abstract collage and painting. Ascott had been interested in ways of representing journeys and territories since a significant encounter with a school supply teacher who wanted him to paint his journey from home to school:

> I thought, where would I start? Would I do a drawing of the house, or the school, or the path? He said, 'No, let me show you what I mean,' and he rolled out this large scroll of paper, and it wasn't about places; it was about feelings, about experiences. It was magic and it changed my life.[63]

Ascott has recently described this teacher's work in more detail, writing that it comprised '15 unfolding panels that constituted a serial mapping of topography and emotion, in which place, passage and perception were woven with a kind of audacity and originality that at 17 I had never before encountered'.[64] Considering this moment of inspiration, it is striking that this teacher's work had an analogical quality to it; it unfolded as a journey unfolds, its form precisely mirroring its message.

It is evident then, that Ascott's practice in the mid-1960s revolved around a balance of process and intuition that he also oriented with the two poles of new and ancient knowledge. Just as the Earth Mysteries movement brought together ancient landscapes with speculations about extra-terrestrial life, the union of *I Ching* and cybernetic theory reflects Ascott's place as part of the countercultural embrace of the mystical and the technological. Shanken wrote that 'For thousands of years the *I Ching* has been consulted on choosing a path towards the future; much more recently cybernetics emerged in part from Wiener's military research, which attempted to anticipate the future behaviour of enemy aircraft'.[65] This insightful comment unites the uneasy alliance of ritual and cybernetics under the banner of prediction. Wiener's early work in the field of cybernetics took place as technology entered into a new arena of possibility, in terms of the anticipation of probability as measured by a plethora of new tools, from surveillance to systems analysis. Predicting – or divining – the future held interesting double implications in the early days of cybernetics, its ancient forms aligned with magic, its modern forms with science. In this light, the map becomes a diagram of possibility, with movement, both hostile and benign, taking place as a constant flow to be measured, analyzed and acted upon, an idea that has significant links to the Ground Control map.

The virtual geographies of Fighter Control

Ascott's spell of national service had made him familiar with the new methods of divining events that were reshaping the military. Technological 'divination' relied on aerial mapping – on surveilling our seas, tracking and following unknown planes. Ascott made a series of wave-inspired works in the 1960s, many of which are now lost, including *Lebesque* (Figure 3.10).[66]

Lebesque and Ascott's other wave paintings and structures depicted the peaked and jagged displays of radar readings; when Ascott was working in bunkers in Newcastle and Edinburgh, they were using Curtain Radar, a by-then slightly out-of-date technology that sent a sheet signal out, which would be disrupted by any contact with a ship or aircraft. Peaks in the on-screen readings represented these disruptions. *Lebesque* is structured around a number of these 'readings', an abstract structure built on echoes of rhythmic flows of information. In this way, Ascott's wave forms were

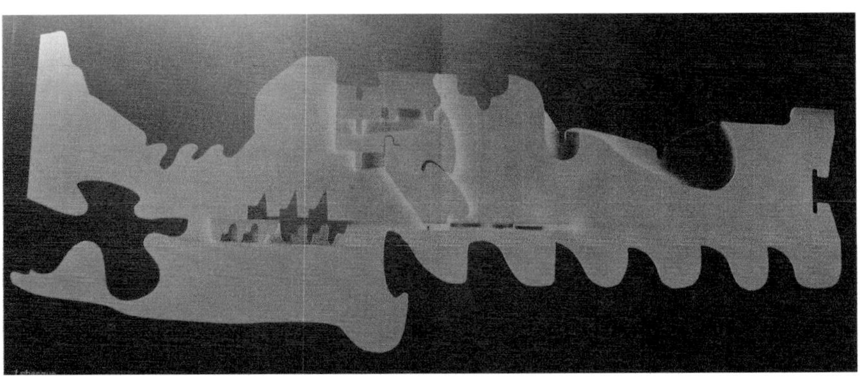

Figure 3.10 Roy Ascott (c. 1967) *Lebesque* 132 × 50 in. Lost. Photograph © Roy Ascott

diagrams of information flow, a signal sent out and coming back as a divination of the new kind. *Lebesque* thrums with the energy of the signal, a static form that simultaneously pulses with a rhythmic rise and fall. In *Inclusion* (Figure 3.11), the tilted hanging of the wave form gives it a fragmented quality that brings to mind broken fuselage or wings.

The layering of the jig-sawed pieces intensifies the fragmentary aesthetic of *Inclusion*. The colour wash is reminiscent of camouflage, a tonal neutrality hovering between earth and water tones. The two very definite curved cuts at the lower right are reminiscent of Ascott's metaforms for growth, having the same sense of double interpretation, inlets or promontories. The wave works are frozen code, signals made into static form. Readings, signals and signs were the currency of the Fighter Control bunkers in which Ascott spent two years, and these signals were eventually displayed in analogue form on the horizontal plane that was the control room map (Figure 3.12). Ascott was one of the many officers that sat in five graduated rows above the map of the North Sea, looking down. It was a bird's eye view, brightly coloured markers pushed into place to indicate the triangulated position of unknown aircraft like a large-scale board game. Ascott recalled that '[...]

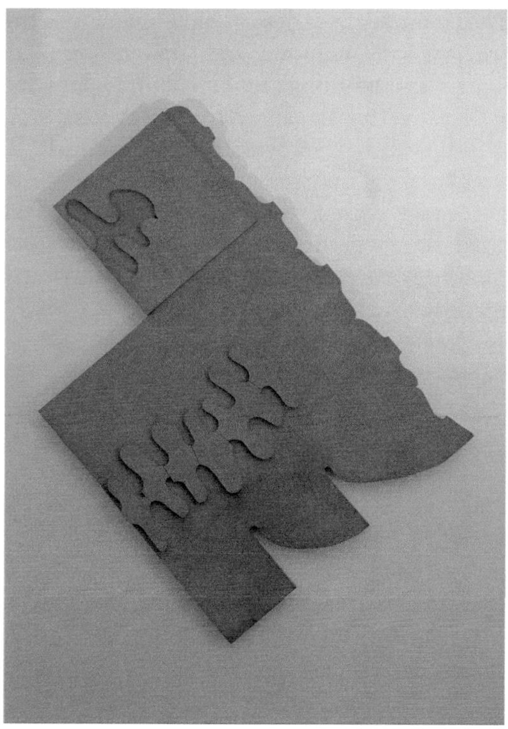

Figure 3.11 Roy Ascott (1966) *Inclusion*. Arts Council Collection. Photograph © Roy Ascott

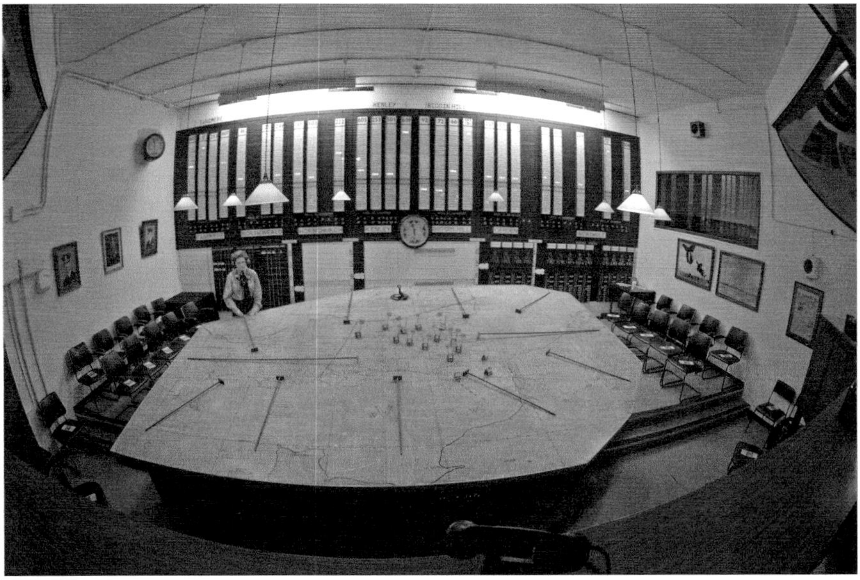

Figure 3.12 Plotting table, Uxbridge. Photograph by Daniel Stirland

you got the job because you were visually adept', an observation that highlights the complex transmissions and transactions of visual, written and aural information that took place in the bunker.[67]

The operation of the bunker was a complex game representing the extended geography of the North Sea, a game played out in codes and viewed remotely, the officers, in their graduated rows, given the god-like power of the aerial view. The large map was laid out in centre stage for the five stories of operatives to view from above. This map became analogous to the situation, mirroring the information that stuttered and blinked its way across the thick glass of circular screens. Radars, monitors, telephones, markers – interconnected pieces creating an analogue map that stood for the unfolding vastness of the North Sea beyond.

In this way Ascott, like so many of his contemporaries, had experienced the most sophisticated technologies of the age in the context of the military, providing in a very real sense, a window on the technological future. He was part of the final generation to undertake compulsory national military service in the UK and he stepped directly from his RAF post into his art school training in 1955. He had applied to be a pilot but purposefully flunked the exam in order to get assigned to Fighter Control, since the training was shorter and a full salary awarded more quickly. Ascott spent his service on the coast of Newcastle and he also served some time at Turnhouse airbase just outside of Edinburgh. We cannot underestimate the powerful, formative experience Ascott had during his time with the RAF. The military bunker was essentially a cybernetic environment; it operated on a principle of systematized man–machine communications.

Along the shores of Great Britain, young officers spent their shifts occupying a curious subterranean world, staggered in layers above a large map of the North Sea, the expanded territory beyond the coast. Gazing at this aerial map viewed from five rows of seats above, the personnel of the bunker were in the then unique position of having access to an extended geography in real time. The radar readings recorded the movements of aircrafts and sea vessels, both known and alien, and transmitted it back to the control room, where Ascott and his colleagues could cross-reference these readings against the authorized flights of the day. The bunker was an impossible entity, the virtual whole bigger than the sum of its material parts. It unfolded space. It was self-contained and hidden from sight and conversely, through its function, it also contained a vast, oceanic geography, made visible through radar screens and translated into analogue form via the map and markers. In this sense, the bunker was a virtual environment. Its dark confines reflected hundreds of square miles. For those who worked in that environment, it offered a glimpse of the technological future; a future in which a computer terminal might collapse the geography of the world, allow us to speak face to face across continents, to communicate regardless of distance.

The bunker was a microcosmic model of the network age that was to come, in which information travels at liquid speed, unhindered by distance. When Jean Baudrillard employed the example of the map in 'Simulacra and Simulation', he created a discourse of the hyperreal which has uncanny parallels with the Ground Control map:

> Abstraction today is no longer that of the map, the double, the mirror or the concept. Simulation is no longer that of a territory, a referential being or a substance. It is the generation by models of a real without origin or reality: a hyperreal. The territory no longer precedes the map, nor survives it. Henceforth, it is the map that precedes the territory – *precession of simulacra* – it is the map that engenders the territory and if we were to revive the fable today, it would be territory whose shreds are slowly rotting across the map.[68]

In Baudrillard's words, it is not the map but the territory that is eroding, the map surviving with its own life. The Ground Control map changed in response to radar readings and the coloured markers that officers would push into place with sweeper sticks. The fixed geography of the paper ocean was superseded by the surface activity of simulacra, there to represent events in real time. The paper map was overthrown. The real mapping took place thanks to the extended vision provided by new technologies, allowing surveillance to take place underground. Baudrillard continued by observing that 'It is the real, and not the map, whose vestiges subsist here and there, in the deserts which are no longer those of the Empire, but our own. *The desert of the real itself.*'[69]

Field theories, then, emerged alongside the disembodiment of information and the dawn of a new global map, which might be updated in real time as our technologies extended our vision and fed constant information from vast territories. Baudrillard's discussion of the map is also relevant to any other analogue information in the age of the digital, in which the stasis of information in printed form renders it meaningless in an age of constant, updating, mutating and accessible dataflow. In this light, the printed form which once signified certainty, finality and longevity has transmogrified into a frozen relic, dead as soon as it is realized. Baudrillard drew out the elasticity of hyperspace, its redoubling quality, producing a reality that '[...] is reproduced from miniaturized units, from matrices, memory banks and command models – and with these it can be reproduced an indefinite number of times'.[70] In this sense, information extends infinitely and becomes dislocated from a concrete, measurable reality; it '[...] is no longer real at all. It is a hyperreal: the product of an irradiating synthesis of combinatory models in a hyperspace without atmosphere.'[71]

In these unbounded geographies of information, objects themselves occupy a peculiar and problematic place. In the military bunker, the map, the markers, the paper sheaves of flight details and the operations clock were analogue, in that they were used to precisely reflect a known quantity or a known piece of information. The radar signals and the blips and pulses on the radar screen were the information flow that provoked the need for new analogue arrangements, such as a bright plastic piece, pushed into place upon the map. This was an analogue game of sorts, played out responsively, the distinction between the printed map and the real territory of the North Sea gradually eroding.[72] For Ascott working as part of this extraordinary live model of information was a prelude for a career in which technologies were possibility – a tool with which to explore the possibility and philosophical ramifications of the new territories of communication that were then just becoming visible.

Ascott recalls the peculiar, singular tension of the bunker and the materiality of the experience of his years in Fighter Control. At this moment of technological history, the network was a visible, physical entity – personnel were wired into headsets, information travelled virtually but was measured by physical objects such as the ops clock, the map marker, the map and the telephone. This phase of development forms an uncanny enactment of the transition towards the systems age for culture, in which objects were absorbed into networks. Journeying back into this evocative, humming environment of military tension shines a light on a process of transition, where the materiality of objects was twinned with the immateriality of information as it flowed through and between the people and objects occupying that underground chamber. It is the perfect example of the humanity of cybernetics; in flight control, the intelligence of living beings was fundamental to the operation and interaction of these advanced new machines.

Ascott never made explicit the links between his art, pedagogy and military service, but he has since realized the extent to which Fighter Control influenced his conceptual and material direction in his early career.[73] Early in his art training at King's College, he produced a drawing that was a 'theatre, full of lights' derived from that wonderfully visual environment he had occupied for the previous two years.[74] If we take into account the theatricality of cybernetic art, from Schöffer's ballet dancing cybernetic sculptures to the immersive environments and theatrics created by the artists and engineers of Experiments in Art and Technology (EAT), then Ascott's youthful sense of performance in the Fighter Control bunker becomes all the more complex.[75] Cybernetic theory articulated the lived experience of the cybernetic age, of military environments and technologies, which were vital to Ascott's work from, and since, the 1950s. It was the living quality that made the military bunker feel theatrical or performative; it was, in a sense, a live cybernetic performance between personnel, and between personnel and machines. We can also identify an interesting disjuncture that will be explored further later in the book, in that for Ascott, his time with the RAF had been positive, formative and fascinating – but for his students, the tension of Cold War politics was to be an unsettling and tense backdrop to their education.

In 1967, Ascott took his work further into the horizontal plane, beginning a series of works that utilized a table top, including *Plastic Transactions* and *Transaction Set*. Originally set up on a gaming table topped with baize, Ascott arranged a series of plastic kitchen implements such as funnels, cutters and pegs. These brightly coloured objects were designed to be held, so they were the ideal pieces for a 'game'. They also had an analogue quality; their forms precisely mirrored their function: a biscuit cutter is the shape it produces. Into the 1970s, Ascott created a number of table-top works of art based upon transactions and interactions.

Shanken writes of this phase of Ascott's work:

> [...] Ascott's experience as a radar officer in the Royal Air Force may have contributed to his artistic predisposition towards a horizontal bird's-eye view and the use of cartographic forms that triangulated information to predict the future.[76]

Plastic Transactions clearly references the working environment of Fighter Control, from the plastic pieces to the grid, looked down upon from above, Ascott remembers it as '[...] an incredible visual experience every way up [...] It's all tied up isn't it?'.[77] It was 'all tied up' within Ascott's practice, but Fighter Control was also a window into the dawning age of communication networks. Ascott has repeatedly emphasized that in the immediate post-war years, it was not yet clear whether the future was digital or analogue, and for a short time it appeared that the technological future would indeed be analogue, not digital.[78] Control rooms were the most sophisticated communication networks of the age, bringing together a mix of analogue and digital technologies, the environment representing analogically the increasingly efficient readings gathered by radar. It is appropriate that it was structured as a theatre with a horizontal stage; the aerial view had transformed warfare. As officers surveyed the scene from the bunker, they gathered intelligence from a position of security and secrecy, above and below simultaneously.

The readymade kitchen implements that were the game pieces were scaled to the hand, designed for handling. They also had an innately analogue visual quality, their form reflecting their functions. A funnel, a biscuit cutter, a peg – as objects, they signify intention, and Ascott played upon this. As a game, *Plastic Transactions* was balanced upon intent and aimlessness, an endless set of possible moves that nonetheless drew its players into thoughtful and absorbed negotiations of form.

In Ascott's *Plastic Transactions* the viewer-participants may take seats in empty chairs and gaze down at the table top, deciding whether to change the arrangement before them in some way. There are no rules. Decisions are entirely personal, and often, if facing another player, reactive; the player assumes a position of conscious control. Ascott's interest in the table-top derived from it being 'the plane of interaction'.[79] While the set-up of a grid with bright-coloured markers bears a visual relationship with the Ground Control map, Ascott's theory of horizontal interaction had evolved from many sources, including the layout of tarot cards upon table tops. On this plane of interaction, the plastic pieces become an evolving map of interaction, never fixed, always capable of transformation.

Horizontality and exchange: information and the surface

Ascott formalized his ideas around the horizontal plane of interaction in a series of works and essays in the 1960s and 1970s. During this period, whilst head of the San Francisco Art Institute, he frequently read the tarot for friends and acquaintances. On one occasion a female colleague had asked advice for her failing marriage; Ascott's marriage was also about to end, so he didn't feel able to advise. He did, however, read the tarot for her from behind the dean's desk. Ascott's brief time there was coloured by an institutional machismo that proved hard to endure, but perhaps his attraction to the extended spaces of paranormal and spiritual speculation offset the pressures of his working life. Laying out the cards on the table was another horizontal exchange, played out like a game, the tarot cards opening an extended world of possibility. He had been further inspired by the Hopi architecture he had encountered while travelling along the Rio Grande. He was struck by the buildings, which joined the horizontal planes of living spaces with external ladders. In the more emphasized movement from the exterior vertical climb in order to enter those interior horizontal spaces, Ascott felt the importance of the horizontal in human communication.[80]

He noted his interest in the 'horizontal element for exchange', adding that 'it does stand quite distinct from the verticality of Western society'.[81] Horizontal exchanges are implicitly non-hierarchical, so Ascott felt that the table top was a democratic space of exchange.

Ascott articulated these ideas in 1975 in a short article called 'Table', in which he argued that the table was 'the medium of exchange, the space between behaviours and states, the grounds for change. A table is very properly a matrix (and so at all levels).'[82] Domestic exchanges of food and drink, conversation and game play all take place on this horizontal plane and as Ascott observed: 'A table exists, just as a house exists and our universe exists, only insofar as it is an arena for behaviours.'[83] For Ascott, the table was an extended metaphor for interaction that expressed the qualities of conceptual art, an art that '[...] aspires to the condition of exchange' and which had '[...] no final outcome, no absolute resolution or irreversible conclusion [...]'.[84] When Ascott began to work on the horizontal, other artists in the UK and the United States were experimenting in the space between painting and sculpture and pressing upon the same issue of verticality and horizontality. This included many of the artists who contributed to Groundcourse in Ealing and Ipswich, such as Bernard and Harold Cohen, R.B. Kitaj and William Green. In the late 1950s, Green had drawn attention for a series of projects in which he created combustible or otherwise self-destructing images. He laid his canvas or board flat on the floor of the studio, using the tyres of a bicycle to spread liquid paraffin and black bitumen into the surface. He was working on the horizontal and the image-creation was a blueprint of destruction, made to be destroyed (Figure 3.13).

It is worth highlighting that collaged and depthless surfaces were manifest in British abstraction – notably by R.B. Kitaj and Adrian Berg before the latter moved into landscape painting. Ascott was also interested in similar trends in American abstraction, including the work of Robert Rauschenberg and Jackson Pollock. Ascott was familiar with Rauschenberg's process of horizontal assemblage and of Leo Steinberg's interpretation of this as 'flatbed' painting by the time he wrote 'Towards a Field Theory for Post-modernist Art'.[85] Pollock and Rauschenberg both attracted formalist interpretations of their decision to work on the horizontal, and both artists were early influences upon Ascott. Pollock's infamous dance of gesture around the floor-laid canvas gave the impression his painting was an analogue chart of his movement, although he always hung the paintings in order to make compositional decisions later in the process; as Steinberg described: 'He lived with the painting in its uprighted state, as with a world confronting his human posture.'[86]

There is power in the aerial view – a power that in this context derives from physical advantage. This power is also more formally implied within maps and charts of territories, including the Control Room map. Steinberg summarized the 'surface distinctions' in the field of paintings as 'a change within painting that changed the relationship between artist and image,

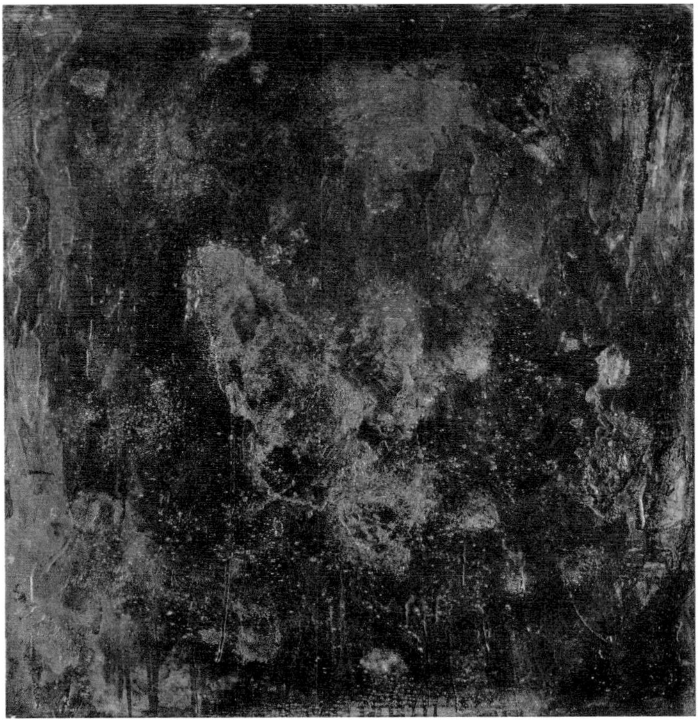

Figure 3.13 William Green (1958) *Untitled* bitumen on fibre board, 917 × 892 × 4 mm, Tate collection

image and viewer'.[87] He saw it as 'part of a shakeup which contaminates all purified categories', bringing Greenberg to mind as he observed that 'The deepening inroads of art into non-art continue to alienate the connoisseur as art defects and departs into strange territories leaving the old stand-by criteria to rule an eroding plain'.[88] This definition has certain parallels with Baudrillard's 'Simulacra and Simulation', particularly in light of the evocative term 'eroding plain', and its relationship to Baudrillard's claim that 'territory' was eroding, its '[...] shreds are slowly rotting across the map'.[89]

In summary then, the aerial map had a discernible transatlantic influence upon the visual arts in the two decades after World War II, connected to overlapping theories of field and territory in critical theory. As I have argued, the analogy of physical spaces or territories for human communication or expression is a trope with a profound relation to technology, and the aerial world-view. As Ascott explored the very human realm of horizontal communication, our vertical reach was expanding, opening out into space. At the same time, communication technologies transformed the ways in which information travelled, from analogue machines or documents to disembodied signals that could be sent instantly. In this way, the concrete map was eroded or even replaced by the flow of signals, of information, that provided a new geography of human movement across the surface of the Earth. When Steinberg argued that although paintings were still displayed vertically they were in essence 'receptor surfaces on which objects are scattered, on which data is entered, on which information may be received, printed, impressed [...]', he implied that working on the horizontal plane evolved from the same concept of information flow and exchange that had occupied Ascott in the early 1960s.[90] Reflecting upon the collisions of science fiction, ancient history and technology in British counterculture, it is evident that these were responses to a shifting geography that allowed us, in stages, to first view the world remotely and to then overthrow the map as information became disembodied. This was a stage in the virtualisation of our world, and the transitions in the visual arts from abstraction, to interaction and then to conceptualism, present a similar process of disembodiment.

Notes

1 For a full discussion of twentieth-century developments of field theory in the sciences, see Donna Haraway (1976) *Crystals, Fabrics and Fields: Metaphors that Shape Embryos*. Berkeley: North Atlantic Books.
2 *Ibid.*, 54.
3 *Ibid.*, 9–10.
4 James J. Gibson (1950) *The Perception of the Visual World*. Boston: Houghton Mifflin; György Kepes (1944) *Language of Vision*. Chicago: Paul Theobald and Co.; and László Moholy-Nagy (1947) *Vision in Motion*. Chicago: Paul Theobald and Co.
5 Kate Sloan *Interview with Roy Ascott* (November 2015) and *In Conversation with Roy Ascott* (Henry Moore Institute, March 2017).
6 Pierre Bourdieu (1993) *The Field of Cultural Production*. Cambridge: Polity Press.
7 Pierre Bourdieu (1984) *Distinction: A Social Critique on the Judgement of Taste*. London: Routledge.
8 Roy Ascott (1980) 'Towards a Field Theory for Post-Modernist Art'. *Leonardo*. Vol. 13. No. 1, 51–2, 51.
9 Roland Barthes (1967) 'The Death of the Author'. In: *Image, Music Text*. (1977) Waukegan: Fontana Press, 142–8.
10 R.B. Brown (2006) [1981] *Katharine Emma Maltwood: Artist. 1878–1961*. Catalogue from an exhibition at The Maltwood Art Museum and Gallery, University of Victoria, June–July 1981. Victoria B.C.: Sono Nis Press.
11 Michelle Foot (2016) 'The Medium and Her Spirit Guide: An Artist's Portrayal of the Psychical Senses'. *HARTS and Minds: The Journal of the Humanities and the Arts*. Vol. 2. No. 3 (Issue 7) 2.
12 Katharine Maltwood (1929) *Glastonbury's Temple of the Stars*. Cambridge: James Clarke and Co., 1.
13 *Ibid.*, 1.
14 Kate Sloan *In Conversation with Roy Ascott* (Henry Moore Institute, Leeds, 8 March 2017).
15 Kate Sloan *Interview with Roy Ascott* (November 2015).
16 *Ibid.*, 2015.
17 *Ibid.*, 2015.

18 *Ibid.*, 2015.
19 John Mitchell (1969) *The View Over Atlantis*. London: Thames and Hudson Ltd.
20 Alfred Watkins (1994) [1925] *The Old Straight Track: Its Mounds, Beacons, Moats, Sites and Mark Stones*. Preston: Abacus.
21 *Ibid.*
22 John Michell (1967) *The Flying Saucer Vision*. London: Thames and Hudson Ltd.
23 John Michell (1967) *The Flying Saucer Vision*. London: Thames and Hudson Ltd. and John Michell (1971) *The City of Revelation*. London: Thames and Hudson Ltd.
24 Bob Rickard (April 2009) 'John Michell: A Modern Man'. *The Fortean Times*.
25 Patrick F. Sheeran (1990) 'The Ideology of Earth Mysteries'. *Journal of Popular Culture*. Vol. 23. Issue 4, 67–73, 69.
26 *Ibid.*, 67.
27 Andy Worthington 'Mystics and Mavericks: The Pagan Reinvention of Avebury'. In: Joanne Parker (ed.) (2009) *Written in Stone: The Cultural Reception of British Prehistoric Monuments*. Newcastle: Cambridge Scholars, 100–10, 102.
28 Nigel Henderson (1953) *Internal Memorandum, ICA*. Tate Archives.
29 Gyorgy Kepes (1956) *The New Landscape in Art and Science*. Chicago: Paul Theobald.
30 The Institute of Noetic Sciences. 'About Us'. Available online at http://www.noetic.org/about/overview, accessed 12 December 2016.
31 Edgar Mitchell. Institute of Noetic Sciences. 'About Us: History'. Available online at http://www.noetic.org/about/history, accessed 12 December 2016.
32 *Ibid.*
33 Kate Sloan *Interview with Roy Ascott* (November 2016).
34 *Ibid.*
35 BBC. 'Renoir: Is that You?' Episode of *Nationwide*. Broadcast on 15 August 1978.
36 Ascott (1980) 139.
37 Pamela Lee (2006) *Chronophobia: On Time and Art in the 1960s*. Cambridge: MIT Press. 7.
38 *Ibid.*, 5–6.
39 Sarah E. Truman (2007) 'Samahdi in Space: An Interview with Apollo 14 Astronaut Dr. Edgar Mitchell'. *Ascent Magazine*. Viewed online November 2016: http://ascentmagazine.com/articles.aspx%3FarticleID=195&page=read&subpage=past&issueID=30.html
40 *Ibid.*, 2007.
41 David Kaiser & W. Patrick McCray (2016) *Groovy Science: Knowledge, Innovation and American Counterculture*. Chicago: University of Chicago Press.
42 This parallel of mind/body and software/hardware was highlighted by Jack Burnham when he prepared the *Software* exhibition for the Jewish Museum in New York, which emphasises its importance to second wave cybernetic art. See Jack Burnham (1970) *Software – Information Technology: Its New Meaning for Art*. Exhibition Catalogue. New York: The Jewish Museum.
43 Arne de Boever *et al.* (2012) *Gilbert Simondon: Being and Technology*. Edinburgh: Edinburgh University Press, 47.
44 Elizabeth Grosz. 'Identity and Individuation: Some Feminist Reflections' in: de Boever *et al* (2012) 46.
45 *Ibid.*, 47.
46 de Boever *et al* (2012) xi.
47 Roy Ascott (ed.) *Technoetic Arts: A Journal of Speculative Research*. Bristol: Intellect Books.

48 Letter from Victor Pasmore. 10 January 1971. Roy Ascott Archival Collection.
49 Richard Hamilton (1955) *Man, Machine and Motion*. Exhibition at the ICA, London.
50 Roy Ascott (2016) 'My Teachers'. *Tate Etc.* April 1, No. 36.
51 Roy Ascott and Kate Sloan. *In Conversation*. Henry Moore Institute, 8 March 2017.
52 Roy Ascott 'Behaviourist Art and the Cybernetic Vision'. In: Ascott & Shanken (2003) *Telematic Embrace: Visionary Theories of Art, Technology and Consciousness*. Berkeley: University of California Press, 152.
53 Kate Sloan *Interview with Roy Ascott* (November 2015).
54 Edward Shanken (2003) 31.
55 John A. Walker (2003) 30.
56 Kate Sloan *Interview with Roy Ascott* (November 2015).
57 Shanken (2003) 32–3.
58 Shanken (2003) 32.
59 Roy Ascott 'Behaviourist Art and the Cybernetic Vision'. In: Ascott & Shanken (2003) *Telematic Embrace: Visionary Theories of Art, Technology and Consciousness*. Berkeley: University of California Press, 152.
60 *Ibid.*, 152.
61 *Ibid.*, 152.
62 *Ibid.*, 152.
63 Roy Ascott (2013) 'Analogical: Incidents in the Life of Roy Ascott'. Issue 127. August.
64 Roy Ascott (2016) 'My Teachers'. *Tate Etc.* April 1, No. 36.
65 Edward Shanken (2003) 33.
66 Ascott recalls that due to his frequent moves around Britain, Canada and the US in the 1960s and 1970s, works of art went missing in action from galleries, schools of art and from the homes of friends.
67 *Ibid.*, 2015.
68 Mark Poster (ed.) Jean Baudrillard (1988) [1981] 'Simulacra and Simulations' *Jean Baudrillard: Selected Writings*. Polity Press. Cambridge,166–84.
69 *Ibid.*, 166.
70 *Ibid.*, 179.
71 *Ibid.*, 179.
72 I discuss some of these issues in the essay 'Analogue Exchanges: Communication Technologies, Surveillance and Selfhood in Roy Ascott's Pedagogy'. In: Christina Albu and Dawna Schulds (2017) *Seeing Others Seeing: Phenomenal Inquiries in Contemporary Art*. New York: Routledge.
73 Kate Sloan *Interview with Roy Ascott* (November 2015).
74 *Ibid.*, 2015.
75 See the chapter 'Game' for a fuller discussion of this issue.
76 Shanken. *Op. Cit.* (2003) 33–4.
77 Kate Sloan *Interview with Roy Ascott* (November 2015).
78 *Ibid.*, (2015) and Kate Sloan *In Conversation with Roy Ascott* (Henry Moore Institute, 8 March 2017).
79 *Ibid.*, 2015.
80 *Ibid.*, 2015.
81 *Ibid.*, 2015.
82 Roy Ascott (1975) 'Table'. In: Ascott & Shanken (2003) 168.
83 *Ibid.*, 168.
84 *Ibid.*, 173.
85 Ascott discussed this in interview (November 2015). Also see Roy Ascott (1980) 'Towards a Field Theory for Post-Modernist Art'. *Leonardo*. Vol. 13. No. 1 (winter) 51–2.

86 Leo Steinberg (1972) 'Reflections on the State of Criticism'. Reproduced in Brandon W. Joseph (ed.) (2002) *Robert Rauschenberg*. Massachusetts: MIT Press. 7–37, 27.
87 *Ibid.*, 36.
88 *Ibid.*, 36.
89 Mark Poster (1988) 166–84.
90 Steinberg (1972) 7.

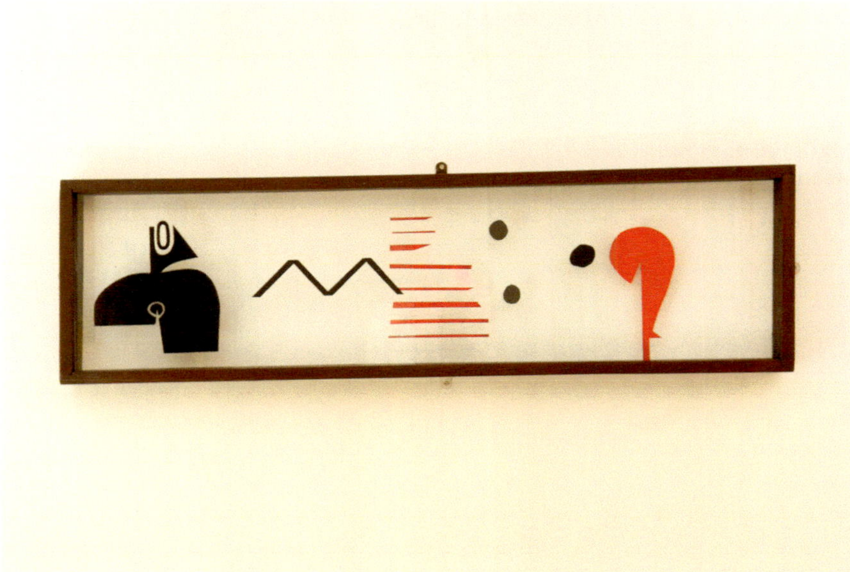

Plate I Roy Ascott (1961) *Change Painting*. Cellulose Paint on Moveable Glass Panels. 66 x 21 in. Photograph courtesy of the artist.

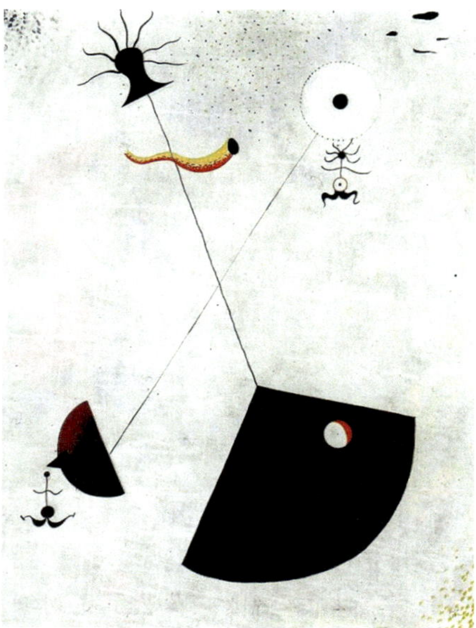

Plate II Joan Miró (1924) *Maternité* oil on canvas, 92.10 x 73.10 cm, National Galleries of Scotland.

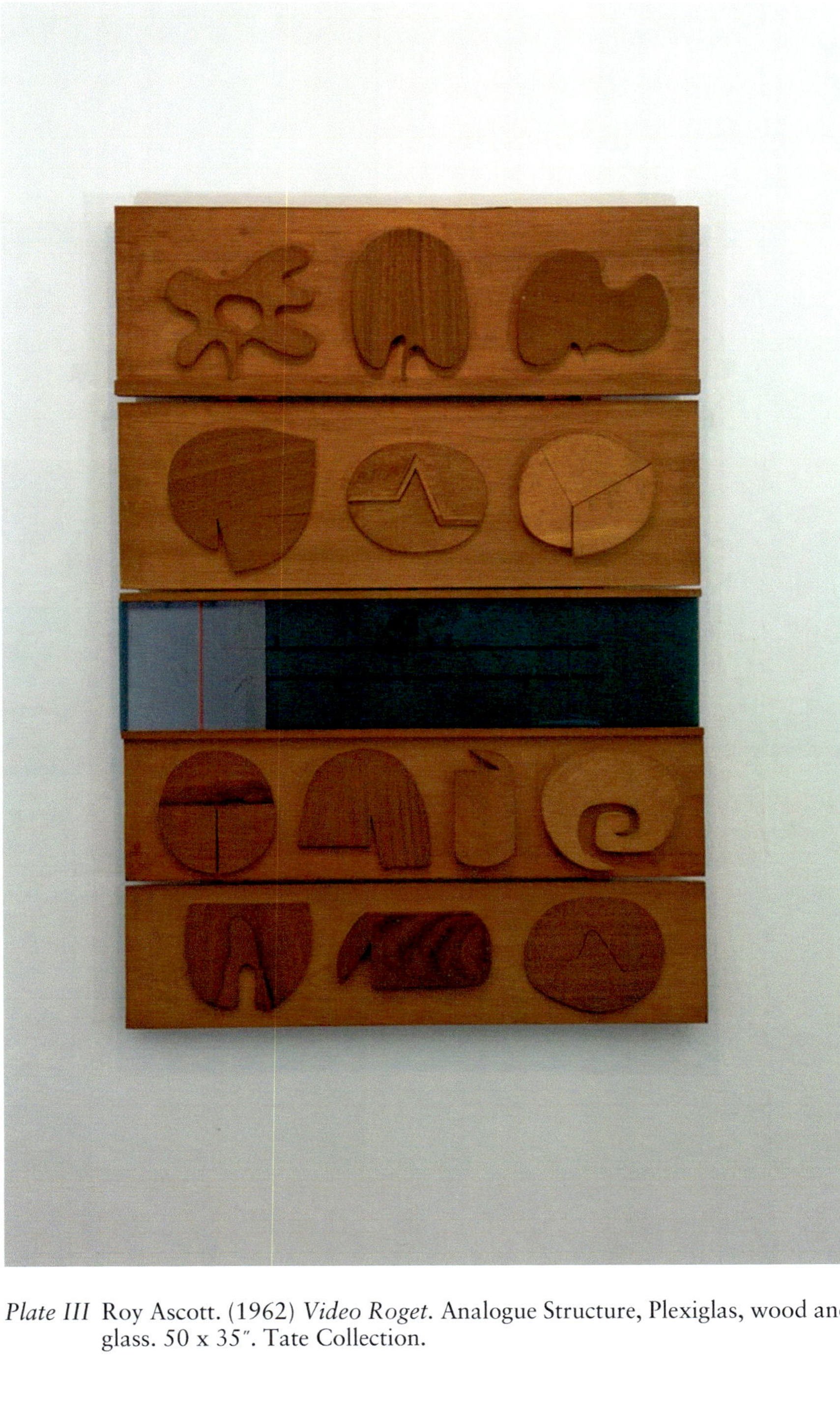

Plate III Roy Ascott. (1962) *Video Roget*. Analogue Structure, Plexiglas, wood and glass. 50 x 35″. Tate Collection.

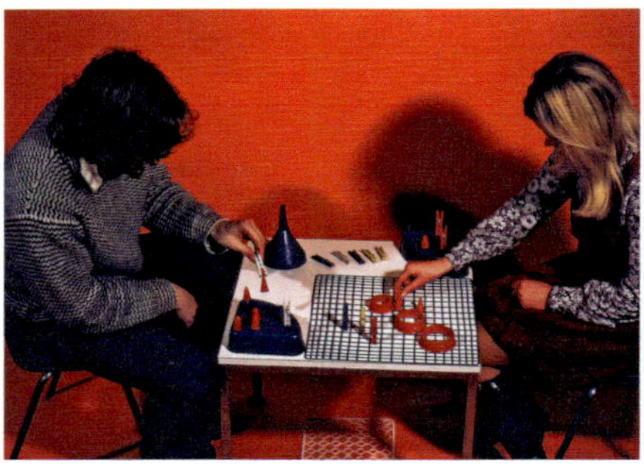

Plate IV Roy Ascott (1971) *Plastic Transactions in Play*. Photograph © Roy Ascott

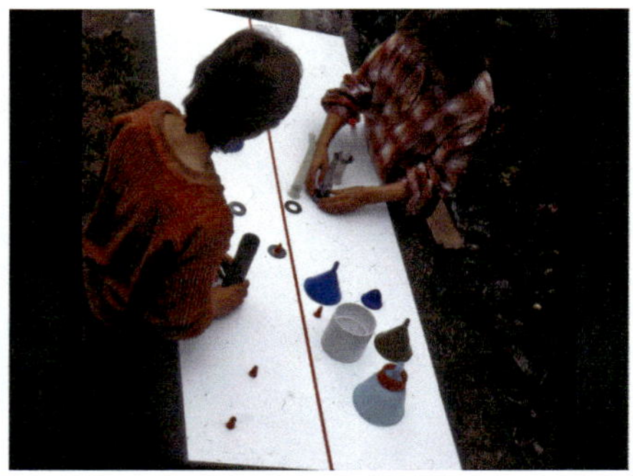

Plate V Roy Ascott (1967) *Detail: Plastic Transactions* Photograph courtesy of the artist

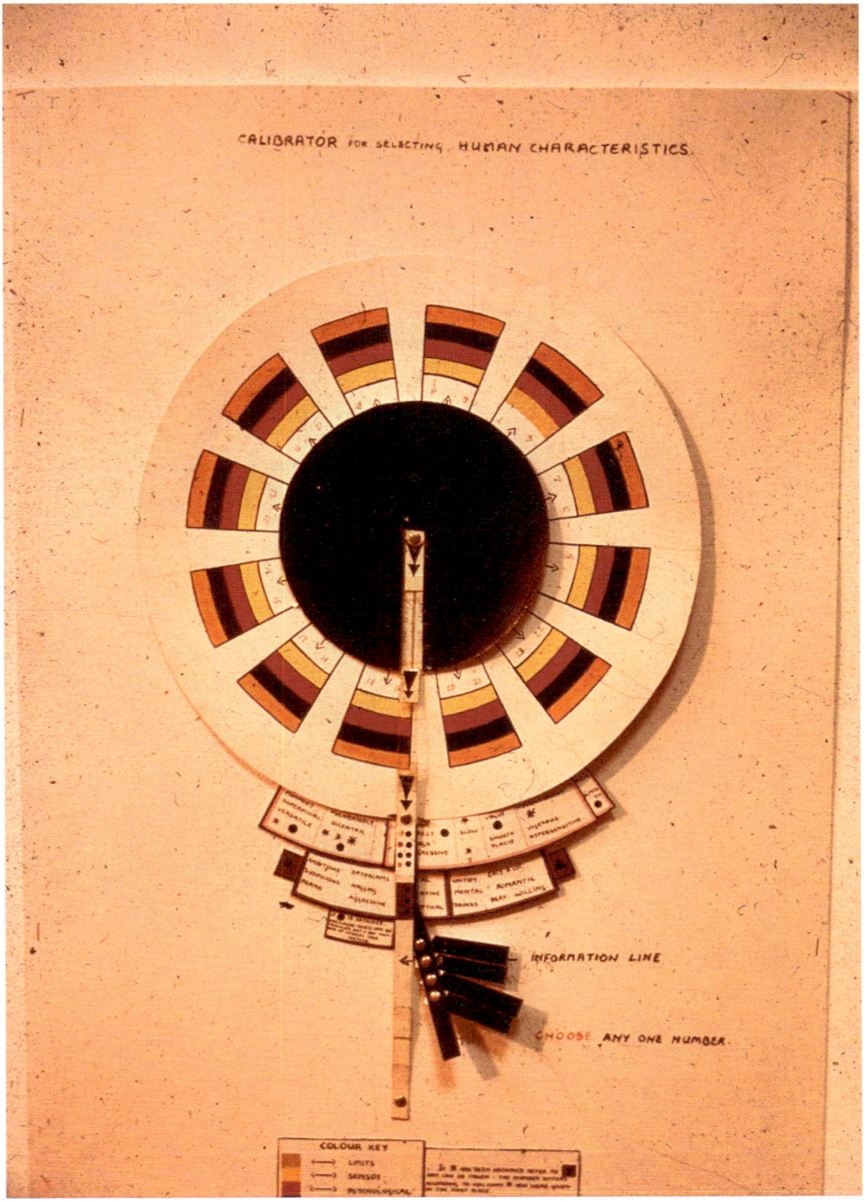

Plate VI Student of Roy Ascott. (1963) *Calibrator*. Groundcourse Ealing. Photograph © Roy Ascott

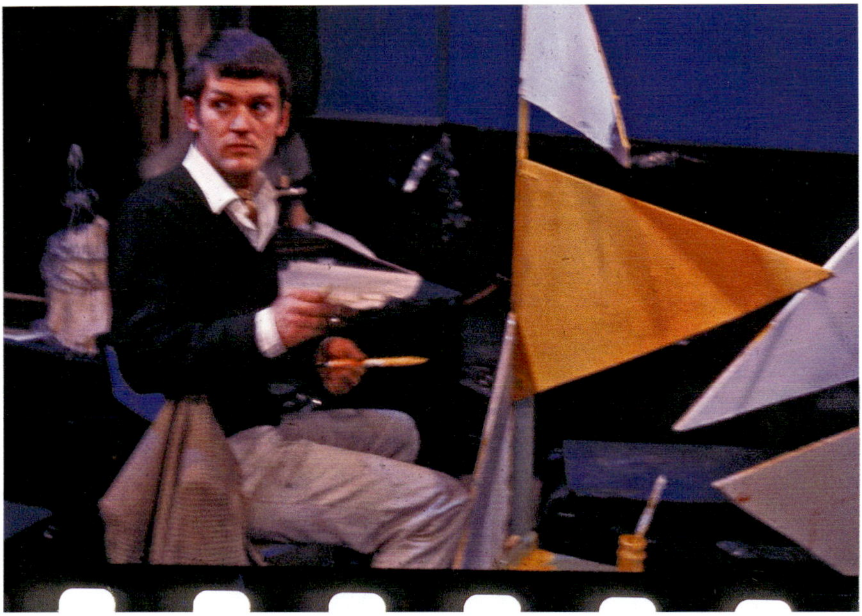

Plate VII Roy Ascott (1963) *Groundcourse Behavioural Project with Calibrator* Photograph © Roy Ascott

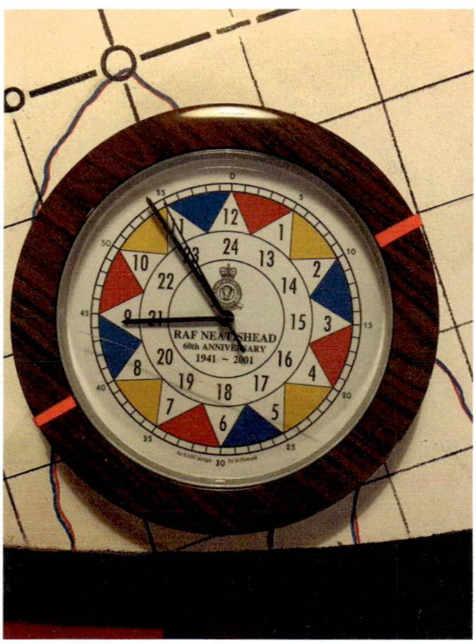

Plate VIII *RAF Control Room Ops Clock*. Photograph © RAF Air Defence Radar Museum, Neatishead, Norfolk

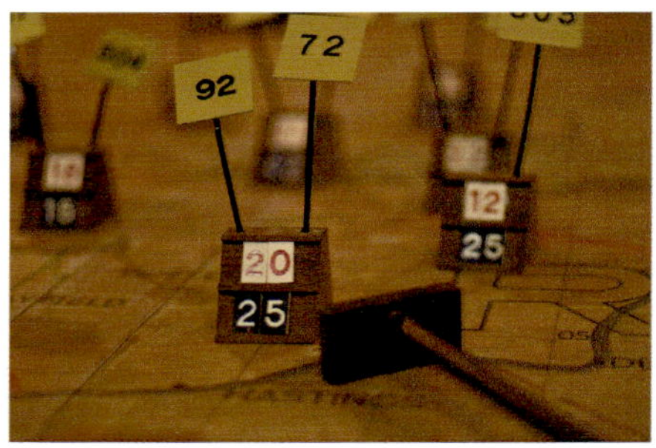

Plate IX Hostile Pilot Marker, Operations Room

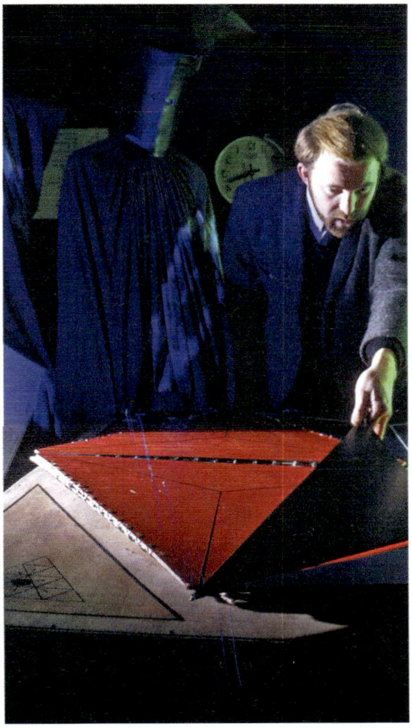

Plate X *Groundcourse Game, Ipswich*. Photograph by Roy Beston, courtesy of Roy Ascott.

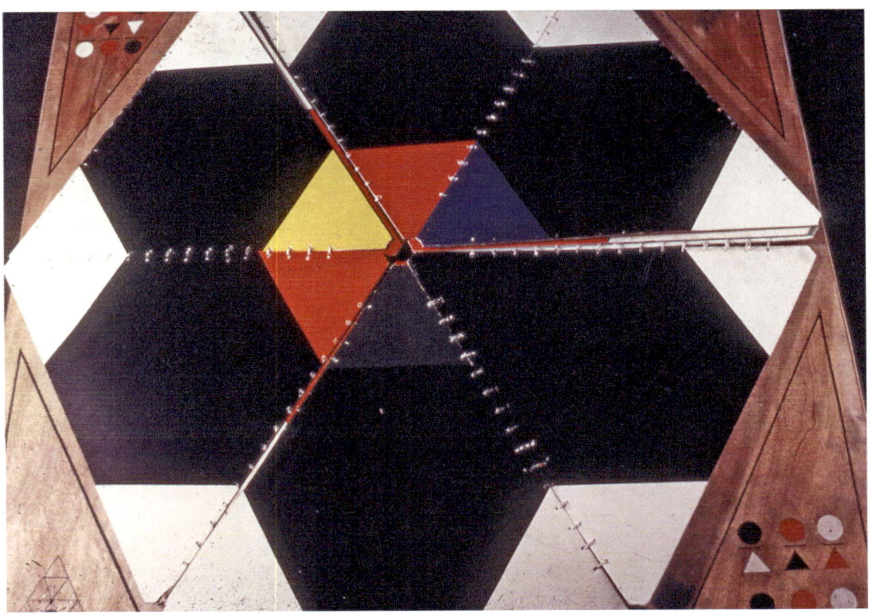

Plate XI *Groundcourse Game, Ipswich*, Photographs by Roy Beston, courtesy of Roy Ascott

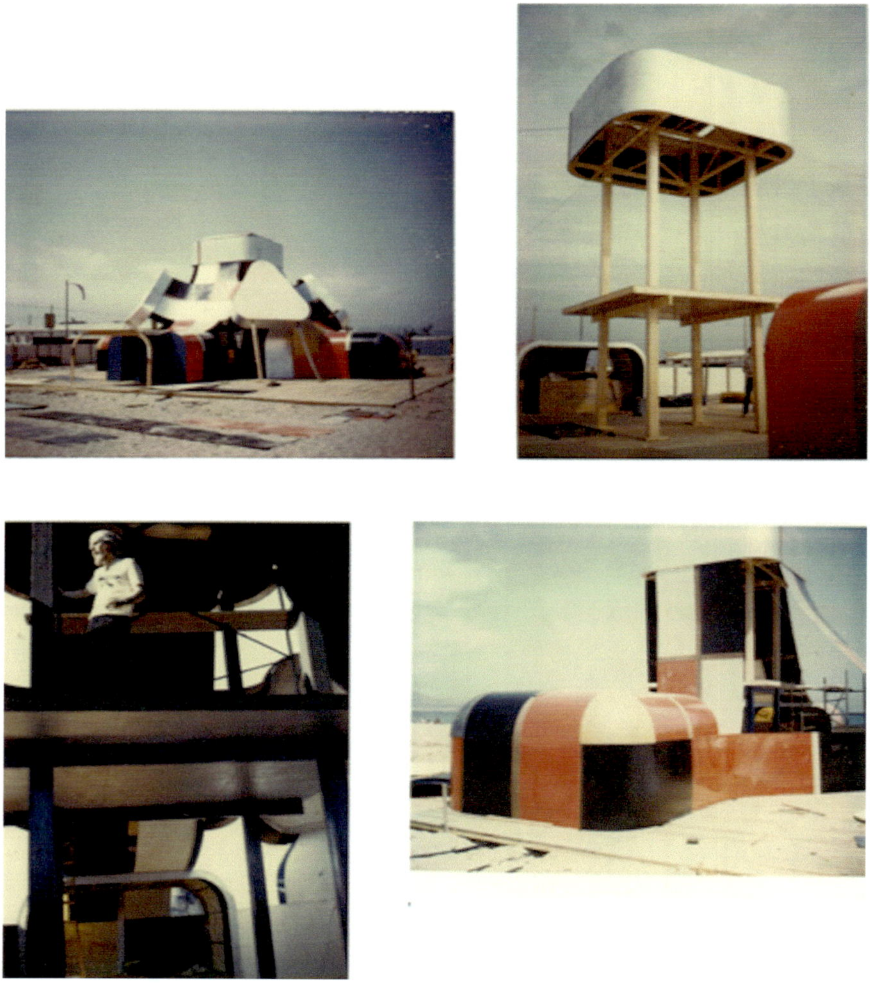

Plate XII Keith Albarn *Discotheque Interplay*. Photograph © Keith Albarn.

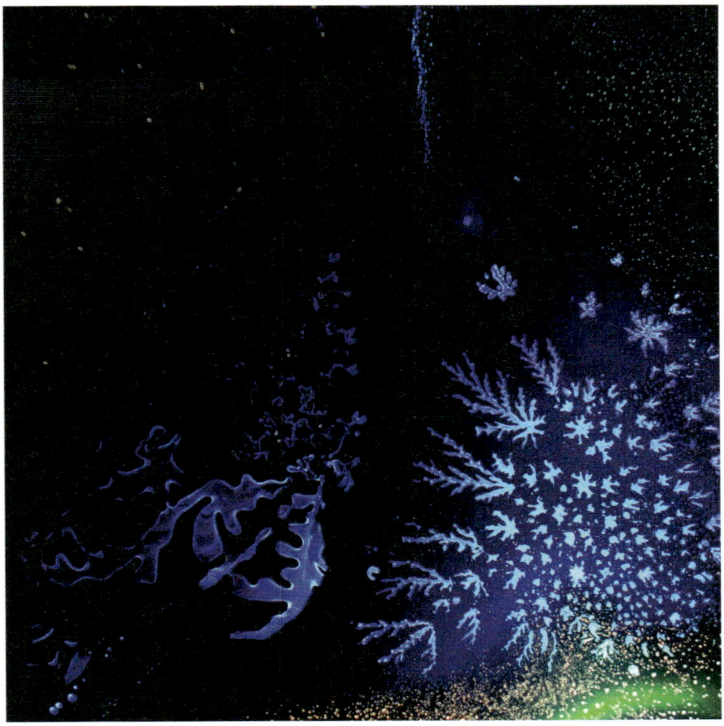

Plate XIII Joel Brown (1966) Slide from *Rebirth*, created for Pink Floyd, performance at the launch of the *International Times*, Roundhouse club. Photograph courtesy of the artist

4 Control

Pedagogy | behaviourism | power

Post-war pedagogy and Groundcourse

Between 1961 and 1967, Roy Ascott developed what was the most radical pedagogical model in the history of British art education. Groundcourse ran first at Ealing College of Art from 1961 to 1964, and then at Ipswich Civic College from 1965 to 1967, where Ascott held the posts of head of Foundation Studies and head of Fine Art respectively. It was a two-year art and design foundation course which brought together cybernetics, behaviourism and collaborative working in what Ascott considered a total system of creative production. As his teacher and mentor Victor Pasmore had recommended him for the post, it had been anticipated that Ascott would create a course around the Basic Design model that he had experienced at King's College. However, Ascott had very recently become interested in cybernetics and his own work had taken new directions as a result. Both incarnations of Groundcourse were immersive, collaborative performances that extended throughout the duration of the courses. Groundcourse stands as the most experimental pedagogical approach to visual arts pedagogy of the century, an approach that used contemporary scientific theory as a framework for re-thinking human visual communication.

Despite this innovation, Ascott had begun a pattern of employment that would dominate his working life over the next decade – introducing radical ideas to much acclaim, before being shown the door for being too radical. His first firing at Ealing by an incoming head rankles in Ascott's memory as a painful and frustrating moment. The new head was very conservative and keen to make his mark – first he banned a visit from David Hockney because it would be 'immoral', and then '[...] he actually fired me! You couldn't get fired in those days [...]. He would clean out the Augean Stables.'[1] This painful rejection was also a measure of the curious conflict raised by Ascott's mode of working, which was undeniably modern and innovative, but equally very difficult to assess. His employer's use of the idiom 'clearing out the Augean Stables' suggested that he found Groundcourse to be a distasteful mess. He also indicated that without Ascott taking the helm, the course would be impossible to run and it therefore needed to be replaced with a more generic model.[2] It had already proved impossible to mark, and the college had not been awarded the right to offer the new Diploma in Art and Design, leaving the first tranche of foundation students high and dry at the end of Groundcourse.

This situation arose from a phase of contradictory conditions in British art education, which were simultaneously liberating and prohibitive. In the late 1950s and throughout the 1960s, art education underwent a series of changes that vacillated between the formal and the radical. While a series of policy changes were implemented that were geared towards formalising art and design training and giving the subjects degree equivalence, simultaneously there was a surge of new and experimental forms of pedagogy.[3] 'The First Report of the National Advisory Council for Art Education', chaired by Sir William Coldstream, had been the grounds of much debate and disagreement amongst artists. During the meetings of the Coldstream board, there had been great debate between members as to how best an art school could function for the future progression of art and design. There was disagreement; factions formed and the report itself was the product of compromise. The Coldstream board had amongst its number the old guard, including Coldstream himself, whose teaching methods followed the traditional structure they had experienced themselves as trainees. However, Victor Pasmore, Richard Hamilton, Harry Thubron and Tom Hudson were also board members, and they argued for a training that would focus on process and material, pushing the strains of Basic Design ideology which they had successfully developed at King's College, University of Durham and Leeds College of Art respectively. Although both Pasmore and Coldstream had previously been members of the Euston Road School, they had differing opinions on how fine art should be taught. During an interview with Lynda Morris in 1985, Coldstream himself commented that 'You should remember that I was only the chairman of the Coldstream Report [...]. It should have been called the Pasmore Report, then people would have understood what was happening.'[4]

Pasmore's vocal contributions saw the introduction of a foundation course for prospective students, as well as a concentrated period of broad study within the Diploma in Art and Design (DipAD) prior to subject specialization. Art and design study was to be placed within the context of a broad, liberal education involving other disciplines, through the introduction of Complementary Studies (akin to General Studies, and not limited or clearly defined by the report), and Art History. Both Complementary Studies and Art History were made into compulsory subjects, which were to be assessed. The idea of the provision of 'a broad context' in the report evidently derives its philosophy from Basic Design with its echoes of the Bauhaus. The Coldstream Report formalized the structure and conditions for the new DipAD and the committee also proposed that 'each art school should be free to construct its own pre-diploma course without reference to any national body'.[5] These factors contributed to creating a short period of pedagogical freedom for British art schools, particularly around 'basic' or foundation years, after the withdrawal of central assessment and before the new strictures of the DipAD. Because basic courses could be devised in-house, they were the natural outlet for experimental approaches. Course organizers had the power to shape a vision of contemporary art practice, an ideology of kinds.

The dance of policy and progress was an awkward one, one partner lagging and the other shooting forward unpredictably. This was certainly the case with the Coldstream Report and the DipAD, which essentially formalized a form of abstract art education that was already in decline. However, its implementation also created new jobs, including Ascott's post at Ealing, due to the nationwide need for staff to teach the new foundation courses that were to be a prerequisite for the DipAD. The foundation courses created a convenient and curious vacuum in a formalized system of art and design education, free from rules. The Coldstream board had imagined the courses would resemble Basic Design, giving students the basic tenets of abstract form with echoes of the Bauhaus. While this sometimes happened, more often the young staff of foundation courses created experimental models that departed from formal abstraction and touched instead upon behaviour, performance or process models of practice. Across London at Hammersmith College, the young Freddy Mercury took part in an art foundation course that frequently employed tactics of visual or psychological projection. A former student recalls being asked to 'imagine you were seeing the room as a house fly', and later, 'imagine you are a junkie for the next two weeks'.[6] In this way, policy enabled chaotic and imaginative progress within the foundation courses, which were at this time more experimental than the DipAD courses for which they were preparing students.

This sudden shift from abstract process to a more immersive, performative and psychologically aware approach to art training is unsurprising when placed in the context of Cold War Britain. One of the most interesting issues that linked art school pedagogies in the period from the 1950s to the 1970s was the prevalence of teaching approaches that employed control systems, surveillance or behaviourism. This was an extraordinary phenomenon, in that strategies of warfare became, for a time, strategies of cultural production. While the long-lasting academic mode of training had also been a controlled system, it had been a system designed to create individual artists that would operate uniquely, maintaining the vital currency of individual genius upon which the art market operated. While the students were part of the same process, all the value was placed in their own finished works of art. Conversely, art training that employed tactical exercises controlling group behaviour focussed instead on process, behaviour and group dynamics – a radical change that reflects the broad influence and infiltration of systems thinking into daily life. It should be noted that the majority of British artist-teachers had experienced either military service or national service and this had a discernible influence upon the development of control-based teaching models and exercises in art schools of the 1950s and 1960s.[7]

Behaviourist systems in the art school

During the 1960s, art students across the UK were the subject of a pedagogical experiment that saw them locked in or otherwise confined, exposed to tactics of shock, disruption, surveillance and made to work within staged conditions or frameworks of limitation. These pedagogical tendencies have yet to be assessed in the context of the pervasive backdrop of Cold War politics and the cultural legacy of World War II. From the end of the war and throughout the 1950s and 1960s, the vast majority of art teaching staff had either served during World War II or had completed national military service after the war, including Ascott himself. The conditions of military service as a controlled collective has an unsettling relationship with the controlled pedagogical experiments of the post-war art school. From locked rooms to collective processes of creation, art education entered into an age of systems of interaction in the post-war years; this is a notable shift for a subject that had a historic focus on individual attainment and development.

This trend towards creative exercises built around group psychology marked a decisive step away from the long shadow of the academy model of art training that had extended well into the interwar years. The most prominent trend in art education after World War II was the use of shock tactics applied to both product and process in abstract education, such as moving or destroying objects of study during the drawing process or ripping, breaking or muddling the students' work. Such interruptions, shocks and disruptions had become an integral facet of British art school pedagogy, often employed as a means of challenging the students in their aesthetic sensibilities. In the opinion of Anton Ehrenzweig, two enduring and conflicting abstract approaches sat at the centre of British art school teaching as they taught abstraction after the war: the disruptive and the constructive.[8] Ehrenzweig wrote about the tactics of disruption employed by Basic Design teaching staff at the Central School of Arts and Crafts, where several artists had taught after the war. He described Alan Davie's attempts to shock the students out of their ideas of 'good' and 'bad' taste in a range of exercises. They moved a chair several times as students drew it, asked them to arrange a set of forms first in good taste and then in a manner designed to insult the teacher, they tore up drawings and instructed students to collage the pieces into a new work of art. This strain of shock tactic hinged on the difficulties of teaching abstraction without students falling into one pattern or habit. Tutors wanted to forge a questioning and interrogative manner in the students and in doing so to give the next generation of artists a degree of intellectual freedom.

However, in 1960s pedagogy, shock and surprise were implemented in more extreme forms, surpassing the more moderate experiments that had found increasing popularity in the immediate post-war years. While there was a certain continuity of thought between the modes of disruption built into Basic Design pedagogy and the approaches taken in the 1960s, there was also a far more fully realized interest in behavioural pedagogical models twinned with surveillance tactics. From the many 'interventions' built into Basic Design approaches, such as tearing up student drawings as part of a project process, to Kardia's locked room, staged systems of control formed the most dominant pedagogical approach of the decade. Ascott's own pedagogical model incorporated his interest in behaviourism within structures of control. Staff scrutiny, that essential element of a visual arts education, was subverted, questioned or made an irritant within systems of shock.

These behaviourist tactics play on the fact that pedagogy is a system of relative control. The word pedagogy derives from a power relationship in that it has an etymological connection to slavery; *pedagogue* derived from the Greek term for a slave-boy accompanying children to school. The pedagogue was a strict, guiding influence that prevented students straying from the righteous road to knowledge.[9] Interacting systems of power, meaning and influence are the fabric of pedagogy. In the twentieth century the word has gained a new currency in educational theory, a catchall term to imply the underlying rules and structures shaping the educative process. The exchange processes that constitute education are defined by relative power and control. Even in today's vast networks of knowledge exchange, in the English language we still define education as 'a process of giving or receiving instruction', with knowledge a commodity that is 'acquired'.[10]

Today both processes of learning and knowledge itself are transforming to the extent that the notion of education as a retained commodity might become extinct in place of virtual, prosthetic knowledge. This vast realm of information is instantly available, challenging the premise of knowledge retention. Given that in the period under discussion, the network age was only on the immediate horizon, Ascott's systemized model of education demonstrated his anticipation of what was to come. He relinquished the concept of education as a commodity of direct exchange between teacher and pupil; instead, he employed systems of control with the intention of fostering student awareness of their own place within broader networks of production. The period of development reviewed within these pages saw the issue of pedagogical control in the art school explored, manipulated and eventually politicized and revolutionized. It is notable that the 1968 art school student demonstrations, which were essentially attempts to seize control and democratise art education, took place after two decades of extremely structured, controlled models of art pedagogy.

Aside from the more extreme controlled exercises, even the structure of criticism and feedback was notoriously aggressive in the 1960s, a fact that corresponds with the increasing complexity of assessing student work after abstraction. 'I've been in that bag, and there's nothing in it', a Hornsey student recalled being told by a tutor upon presenting his painting at a group criticism, a typically casual dismissal.[11] Ascott's course subverted, assessed or attacked this model of judgement on several levels, something that is confirmed by his grade policy. You might pass or fail; complete the course and you would pass. There would be no grade bands. As one of the first art courses to be almost entirely concerned with process, the students left the course with a scant collection of outcomes, far from the portfolios of drawings that were typically used to assess students. Instead, the students amassed diagrams, rudimentary machines and the detritus of game play or performance. In this way, Groundcourse abandoned outcome in favour of process, valuing participation in the course as a form of collective action without singling out individual attainment.

While this approach to assessment was essentially egalitarian, it was also a form of social control in that students could not choose to operate in isolation, forced to abandon solitary concentration and to participate instead in collective action. In this way Groundcourse was one of the many control structures employed within British art school studios in the post-war period, when behaviourism, interactivity and group dynamics formed a vital and telling counterpoint to the ingrained individualization of creative process in art education. This was a transitional time in the school of art, in which new approaches were produced in relation to the decline of abstraction and of modernism, the move from object to concept, from individual to network. Ascott's model of pedagogy was an experiment that set up possibilities that led to unpredictable or uneven outcomes, at times failure but often great invention, creativity and complexity. Ascott's conception of the course was that of a total system in which students and staff reacted to the material and environmental limitations and possibilities around them. This organic–cybernetic model treats an organization or group activity as a networked whole, examining the role of specific groups and individuals in order to assess efficiency and maximize potential.

Behaviourism and Groundcourse

Ascott designed the underlying behaviourism of Groundcourse on the principle that systems were alive; he presented his integrated pedagogy as an organism in itself. The interactive element of the pedagogy was a key principle for Ascott, who recalls:

> Cooperation, participation was important [...]. The next thing was let's design a machine and build it, a machine that would enable you to relate to what's happening out there to what you do – the environment and your behaviour that you could use. So what would be the variables? They'd go to their mind map and the mind map would say feeling is important, logic is important whatever way they saw the world.[12]

Student cooperation was important for the very reason that Ascott viewed the course in its totality as an organism, so every person, activity or object played a part in its total function. If we take this in parallel with Ascott's concept of the 'analogue' sculpture or structure, the course in itself enacted

Actor Network Theory as creative process. In this model of operation, each student or member of staff had a part to play, as did the devices, games, drawings and objects they created. The outcome was that of total function, abandoning the idea that the objects and drawings made by the students were the end-point of a process. Rather, they were both produced by the total function of the 'organism' and then these created objects became contributors to that same function.

There are a few important points to underline here with regards to systems approaches to pedagogy. Firstly, the notion of networked, participatory function was a new development in post-war society, and the same period saw the development and growth of both anthropological concepts of cultural evolution as well as the theoretical area of 'organizational science'. Stafford Beer's early cybernetic models for organizational management played a key role in forging the discipline. In *Brain of the Firm* Beer evolved a business management model called the Viable System Model (VSM) based on the human brain. Beer's VSM could be used to organize entities (people, machines, money) and position them within a firm in order to optimise the production of goods or services. The various 'levels' corresponded to functions within the human brain. Stafford Beer's *Brain of the Firm* later became an important point of reference for Brian Eno, something that will be further explored later in this book.[13]

For now, it is worth highlighting that using the biological function of the human brain as a model for organizational function was part of a wider trend in British cybernetics, which Andrew Pickering has interpreted as a vital influence upon 1960s counterculture.[14] Beer's book was a crystallisation of two decades of cybernetics, tying to the trend in Britain towards modelling and understanding the function of the human brain. From Ashby's *Design for a Brain* to Beer's schematized brain as a model for an organization hierarchy, it appears that the human brain held an emblematic place within the conglomerating systems research of the post-war years.[15] It was the ultimate black box: a complex mechanism that could be better understood through forms of experiment, including behaviour modification. In Pickering's account of the links between cybernetic models of psychiatry and mind control and the mind-expanding pursuits of psychedelic culture, we can also read that curious relationship between scientific invention and the purely creative use of technology. Both are experimental and both produce unanticipated results.

Ascott and interdisciplinary systems research

During the 1960s, Ascott assimilated literature from different fields that presented research on systems models of social behaviour, education and control. He read J.H. Clark's 'Adaptive Machines in Psychiatry', in which the author assessed the potential of new learning machines for psychiatric assessment, therapy or personality testing.[16] This article was illustrated with network diagrams that created feedback relationships between depicting the flow of information between instructor, pupil, the 'task' and its comparator, as well as a simple feedback loop of patient, psychiatrist and a control mechanism representing behaviour as an adaptable quantity.[17] In the essay Clark used the term 'calibration', which was to be such a key idea for Ascott, to describe the process by which variable behaviour could be controlled through psychiatric intervention. The author also mentioned Pask's teaching machine SAKI, as well as Skinner's teaching machines.[18] Skinner had used a variety of machines to record operant responses, such as disks or levers that could be pushed by rats or pecked by pigeons. Skinner drew a comparison between a rat operating a lever to receive food and a human operating a lever on a vending machine to receive candy, each action brought about as a conditioned response.[19]

By the early 1960s, diagrammatized networks were beginning to make their appearance in pedagogical contexts, not only in relation to machine learning. In Ascott's archive of the 1960s, there are copies of Open University curriculum diagrams, articles on systems education and letters from other artists working with similar ideas across Britain. Anthony Blake wrote to Ascott in 1964 to describe his teaching at Kingston College of Technology, where they were '[...] using the principles of systematics for two complementary purposes'.[20] Blake was a research fellow at the Institute for the Comparative Study of History, Philosophy & the Sciences.[21] He described the two ways of using what he called 'systematics' as follows:

> The first is to derive structures for the themes which we discuss in the groups, and which is the basis for the production of ideas and written work by the group. The second is to derive a structure for the group process itself. There is a third line of work connected with explicit teaching of psychology, which is something we specialise in at the institute.[22]

In this interdisciplinary department of Kingston College of Technology, some of the same principles of collective organization and function that Ascott had employed since 1961 were built into the course. Later in the same letter, Blake professes his intention to work towards '[...] establishing connections with educational automation [...] one of our main themes is the creative use of modern technology'.[23] At Kingston then, the curriculum employed systemic structures that shaped interaction and defined the collective direction of work, while simultaneously teaching about psychology and communication technologies. Blake was also in discussions about a potential partnership with the firm De la Rue Bull, a collaborative venture in machine design between De la Rue and CMB, emphasising the growing interest in the future prospects of the machine as a learning device. Pask's *Eucrates Training Machine* and *SAKI* (self-adapting keyboard) had been at the cutting edge of this new realm of engineering in the 1950s. These machines were designed that responded to a user's learning in a progressive way, testing and challenging the user as they learned, for example, how to type with speed and accuracy. Beer later ruminated that the original cybernetic harmony had been lost through engineering, that:

> […] they saw themselves as designing a machine to achieve the content-objective (learn to type) instead of building a Paskian machine to achieve the cybernetic objective itself – to integrate the observer and the machine into a homeostatic whole.[24]

This was to some extent a comment upon the dissonance between the purpose-built machine and the experimental and unpredictable nature of the earlier cybernetic machines developed in the UK by Pask, Ashby and Walter. As a personal friend of Ascott's, Pask was a source of both inspiration and information. Pask gave a talk to Groundcourse staff and students at Ealing, and like Ascott he was a member of Cedric Price's and Joan Littlewood's *Fun Palace* committee, a cybernetic architecture project that was never realized.[25]

In the 1960s, Ascott organized, made contributions to and participated in several research events pertaining to cybernetics and systems thinking, including a number that took place in art schools. In 1968 Ascott was invited to submit a paper to the Institute of Computer Sciences for an 'International Conference on Cybernetics' planned for the following year, which had been divided into several subsections. The sections were together intended to form a debate of the meaning of the term cybernetics in relation to neurology and biology, industry (automation), social and economic consequences (including management), artefacts, the natural sciences and the social sciences.[26] He talked at the conference 'Creativity in Art and Science' which took place at a Polytechnic College in London in 1969, as well as at a 'Systems Symposium' at Birmingham College of Art and Design the same year.[27] Ascott had planned a symposium titled 'Art, Science and Behaviour' at Ipswich College of Art in 1964, which unfortunately fell through. He had secured a number of speakers for the event, including Pask, Frank Popper, Frank Melina and Gustav Metzger. The cybernetic sculptor Nicolas Schöffer had also submitted a paper.[28] Ascott wrote in his planning notes that 'science and technology have had a great impact upon the plastic arts in this century. Scientists, artists and commentators are eager to strengthen links between the two cultures.' He proposed that this symposium would be a chance for them '[…] to meet and discuss their ideas and attitudes'.[29]

Had this event taken place, it would have brought together many of the leading voices in the interdisciplinary field of cybernetics, including scientists, artists and theorists. Ascott himself was a member of the *International Congress of Cybernetics*, which met in London in 1969 and discussed its aims of establishing both a 'World Union of Cybernetics' and an 'International Cybernetics Foundation'.[30] Over the course of the 1960s cybernetics moved from the alternative to the mainstream, and Jasia Reichardt's 1968 exhibition *Cybernetic Serendipity* at the ICA has become a dominating reference in the history of cybernetic art in the UK. However, it is evident from even a brief assessment of archival material that important work on systems thinking, interdisciplinary partnerships between engineering firms and art schools, and more informal friendships and communications were taking place throughout the decade.

Ascott's weaving of behaviourist and cybernetic theories in Groundcourse played upon the interrelationship of the two fields. The complexity, the unknowability and the intricacy of the human mind were problematized in both fields, creating a field of inquiry that dealt objectively with the most subjective problem of all: personhood. In a world that every day brought forth new machines, devices and applications, the human subject remained our greatest mystery. This was the age in which human potential became relational, judged in contexts of application that also included machines. We might see behaviourism as a strain of distanced control and scrutiny that started from the assumption that human behaviour influenced bigger systems of operation: military, economic, social and cultural. Vitally, any critical discoveries around behaviour control, systems of behaviour or modelling behavioural responses were dependent upon scrutiny. There is an interesting parallel then between the psychological tension of scrutiny and the tense surveillance which characterized the Cold War. Both behaviourism in itself and its application within Groundcourse referenced the visual language and physical interactions of what could be described as cybernetic warfare.

Theorising behaviourism: *Control Magazine*

During Ascott's tenure at Ipswich School of Art, a number of the artists involved contributed to *Control Magazine*, a journal that was founded by the artist Stephen Willats. Willats was an alumnus of Groundcourse Ealing. He started *Control Magazine* after a 'lunchtime discussion' at Ipswich College of Art, where Ascott had invited him to teach after his studies at Ealing.[31] During this discussion, some of Willats' colleagues had pointed out the lack of publications that recognized what was happening in the visual arts in the UK, a problem Willats hoped to address in founding a new magazine that homed in on critical issues. In the editorial for the first issue, Willats wrote:

> Control's main function will be to publish articles by the personalities which make up the new attitude in visual communication. Control will be organic in the sense that each issue will either be given over to a group of people which present a unified point, or will deal with a specific subject, and various designers etc. with different approaches will be asked to contribute to an issue, this will insure that Control becomes fluid, and also vital in so far as it acts as a common forum.[32]

Willat's intentions recall Ascott's own application of the principle of 'requisite variety' in hiring staff and programming exercises for Groundcourse. *Control Magazine*'s contributors were Logie Barrow, Mark Boyle, Dean Bradley, Willats and Ascott. The title was reflective of their desire to formally explore the cybernetic and behavioural tendencies they had developed within fine art pedagogy and practice. Willats also wrote:

> The magazine is intended to function as an agent of reform within the preoccupations of the art community [...] The theoretical foundations of art will need to be self-determined by the artist, rather than being left, as in the past, to others. For example: the Art Historian.[33]

These words anticipated what was to be significant change, both in how artists are trained and the expectations placed upon artists by funding bodies, galleries and other institutions. To self-theorize is indeed vital to contemporary practice, to the extent that work often has to be clearly theorized before it is completed in order to secure funding or other opportunities.[34]

In 1965, the year of the first edition of *Control Magazine*, Willats conceived of the idea of the 'Conceptual Designer' for his own practice, a phrase he intended to express the role of the artist in what Willats saw to be '[...] a more complex, fluid and positive vision of a new society'.[35] Over the course of the second half of the decade, Willats continued to create kinetic and light art, while becoming progressively interested in a more conceptual form of practice. By the start of the 1970s he had begun to employ social science methodologies such as questionnaires for a select group of participants in order to create work that networked, or diagrammatized, their habits and beliefs. *The Social Resource Project for Tennis Clubs* (1972), *The West London Social Resource Project* (1972) and *The Edinburgh Social Model Construction Project* (1973) all involved what was essentially a loose form of social network analysis.

Willats' work often took the form of refined diagrams extracted from those produced during the projects. John A. Walker has observed that at the time, the work divided opinion. For example, after seeing Willats' work at the 1972 Serpentine exhibition *Art for Whom*, the artist Faure Walker wrote that it was '[...] condescending and manipulative, experimenting on the subjects through maze-like either/or recognition tests as if they were rats learning to press Pavlovian stimulus-leavers'.[36] The discomfort provoked by Willats' work hinged upon whether or not the viewer considered his social interventions empowering, through the provocation of self-analysis and subsequent self-knowledge, or degrading, in that the subjects were operating within a predetermined system, like rats in a maze. The issue was that a controlled system is one of relative power. For Faure Walker, the darker implications of behavioural science did not sit comfortably as an approach to social art practice.

John A. Walker also mentioned the 'rather sinister' title of Willats' magazine as evidence of his 'cold and scientistic' art.[37] It is, however, important to acknowledge that the artists working with cybernetic and systems theories in the 1960s often did so with a fundamentally democratic approach. A system of art production reduces the power of the individual creator; instead, it disperses process, meaning and reception as subjective elements of a far larger system. Faure Walker's criticisms came in the early 1970s, by which time hierarchical or state-driven models of social control sat firmly in contradiction of countercultural politics. During the techno-optimism of the 1960s, however, the requisition of ideas from the fields of behaviourism and cybernetics had yet to develop the 'sinister' implications that John A. Walker described retrospectively.[38]

In the UK as well as the US, institutions had fostered collaboration and partnership between art, engineering and science.[39] During this time, as Matthew Wisnioski has noted, '[...] art took on a range of shifting meanings amongst scientists and engineers that helped construct the very notion of the "technological arts"'.[40] He also recorded the decline of this newly minted area of practice by the start of the 1970s in the US, after the civil unrest that had seen out the 1960s and against the backdrop of '[...] an economic recession in the military-industrial complex'.[41] While the economic situation in the UK had been very different to the US since the end of World War II, the same political unrest extended across the globe and distrust of technology formed an element of that unrest. For those artists who were forging the path for media art, technological advancement was an extended realm of communication and creativity for artists. This went beyond the use of machines and into structures and frameworks of human communication, informed by the systems theories that were evolving rapidly alongside contemporary science and engineering.

In 1966, Ascott contributed a statement to *Control Magazine*. The cover of the early journals each featured a plain circle in colour – Ascott's statement echoed this form in its concrete arrangement of text. The circle structure was an exercise in formal control, forcing the prose into shape. In his statement for *Control*, Ascott articulated a creative control system of reciprocity, where each element in a system had influence upon each other in a manner that exacted change. He wrote: 'In art the will to control is expressed through processes of restricting experience and of creating in familiar relationships within a universe of visual discourse.' He then wrote: 'Just as my own artwork feeds back to affect my subsequent behaviour, so in society generally the artist activity may function as some kind of ritual control mechanism. Both individual artworks and cultural clumps can act as behavioural triggers.'[42]

Ascott argued that the artist controls the processes and reactions around the work of art, but also that this was not a sealed, complete form of control, not steady but rather '[...] affecting a changing, fluid field'.[43] This is a more open philosophy of control in which the artist is a participant within a system rather than in leadership of it, something that he also underlines in his observation that '[...] my own artwork feeds back to affect my subsequent behaviour'.[44] In this sense, Ascott's 'Control Statement' placed artists and their creations in much the same light as cyberneticists and their experiments – not a creator and observer, but part of the system even as it performs its function. As Pickering has observed:

> Second-order cybernetics, that is, seeks to recognize that the scientific observer is part of the system to be studied, and this in turn leads to a recognition that the observer is *situated* and sees the worked from a certain perspective, rather than achieving a detached and omniscient 'view from nowhere'.[45]

Pickering's performative ontology is equally relevant to the more situated art practices that were emerging simultaneously, many of them directly informed by developments in technology.

The energetic core: Groundcourse Ipswich

When Ascott took up a post in Ipswich, he revisited his curriculum design, reconceptualising it as 'the energetic core'. In a sketchbook from 1964, Ascott drew and noted ideas for his pedagogical approach for a prospectus design, most notably a circular curriculum design that recalls Walter Gropius' famous Bauhaus structure. In this diagram (Figure 4.1) Ascott sketched concentric circles around an 'energetic core', each outer layer adding detail. The diagram lists a core of proposed teaching staff in one layer, including Stroud Cornock, Anthony Benjamin, Tom Phillips, Ron Kitaj, Colin Moss, Steve Willats, Geoffrey Clarke and Ascott himself. The outer ring lists the various elements of practice and theory that would inform the energetic field of the course: 'paratheatre, kinetics, cybernetics, photoworks, new sounds, cine, behavioural systems analysis, sign design, data handling, painting, sculpture, construction'. Although he had been ousted from one art school and taken up residence in the other, Ascott had not strayed from his vision of an interconnected field of communication and art. Instead he was thinking of new ways to describe this approach in visual and written form. While both incarnations of Groundcourse prompted the production of exciting work, by the time the Ipswich course started it is evident that Ascott had ideas for a more integrated approach in which artist-designers would work with the technologies of film, theatre and more kinetic-mechanical performance. One of the notable elements of this curriculum structure is his two strands of 'fine art' and 'communication design' as the twin poles that made up a Groundcourse.

Elsewhere in the same sequence of sketchbook notes, Ascott mentioned the need for 'audio-visual media', and lists CCTV, sonosound and a projector amongst other proposed equipment. In each of Ascott's teaching posts over the following decade, light studios were essential elements of his teaching facilities and this was certainly his ambition for Ipswich. As Ascott

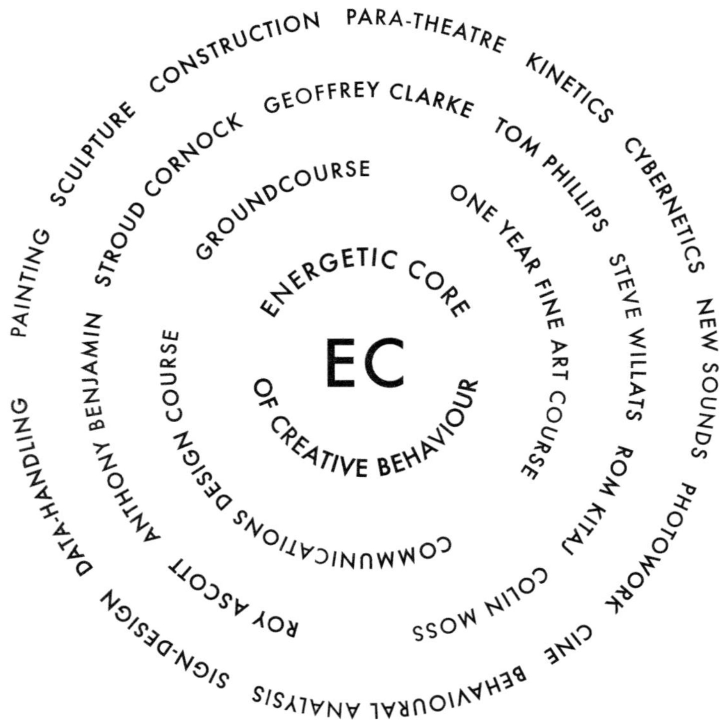

Figure 4.1 Roy Ascott (1964) *Energetic Core Curriculum Design for Ipswich.* Diagram created from a sketch by Arabella Harvey

moved on to the framing concept of an energetic core, multimedia became a more central part of the Groundcourse vision. Indeed, the vision that had inspired Groundcourse in the beginning – that of an interactive, technology-driven future for visual communication – was moving into reality at pace. In a decade of light shows, concept-rock concerts, cybernetic para-theatre and art/science/engineering collaboration, a new field of practice and discourse was certainly emerging. Amongst this tangle of equipment detailed by Ascott, the presence of CCTV is of particular note – simultaneously recording and playing, it was a technology of the instant that was also symptomatic of an increasingly surveillance-oriented culture. It is for these reasons that the photographic records and recollections concerning Groundcourse Ipswich have a decidedly theatrical character. As the Ealing course ended, many of the students left with an anticipatory sense of the futuristic potential of technology and cybernetics, as well as with glimpsed visions or dreams for future projects that would be multimedia, interactive and sensory. However, as the Ipswich course began, these ideas were less wildly radical, while remaining avant-garde. In the space of a few years, the potential for multimedia approaches to interactivity in the visual arts had only grown.

When these controlled approaches were applied to visual arts pedagogy, a different set of problems arose. The question of relative power in education is a source of continuous debate and the perceived imbalance of power has often been the stimulus for revolution. Indeed, control within the context of education became a politicized issue over the course of the 1960s, culminating in the worldwide protests of 1968 and articulated simultaneously by the work of Paulo Freire and Henri Giroux in the form of Critical Pedagogy.[46] Critical Pedagogy was a new branch of pedagogic enquiry that sought to question the way that educational institutions control and manipulate knowledge-as-power, particularly within the context of wider economic and political agendas. Henri Giroux's 1991 article *Border Pedagogy and the Politics of Modernism/Postmodernism* advanced an open field of pedagogy that allowed for reconfigurations of meaning.[47] Giroux writes:

> [...] I advance the most transformative aspects of this version of critical pedagogy by articulating a theory of what I call a border pedagogy of postmodern resistance. In this perspective, the issue of critical pedagogy is located within those broader cultural and political considerations that are beginning to redefine our traditional view of community, language, space, and possibility. In short, border pedagogy acknowledges the shifting borders that undermine and reterritorialise configurations of culture, power, and knowledge, and links pedagogy to a more substantive struggle for a democratic society.[48]

This description of a pedagogy built upon shifting and connected social, cultural and economic systems echoes a broader interdisciplinary concern with the issue of connectivity in postmodernity – the terrains we negotiate and inhabit, on physical, psychological and cultural levels. This drew in cultural studies, sociology, organizational science and the arts.

Ascott intended to free the students through giving them control over process. While they were part of a defined environment of systematized production, this was a system designed with open possibility in mind. At times though, the course strayed onto uncomfortable territory; the student body was subject to controlled experiments; their modes of behaviour were scrutinized and adjusted, their preconceptions dismissed and replaced, their own agency limited by their place within a larger 'organism'. The organism of the student body was to be fluid and reactive but it was also designed, and it had perimeters. For Ascott though, control in the context of arts pedagogy was about creating structures for human behaviour, about organization and about the pursuit of the most productive systems possible. Ascott's statement in *Control Magazine* ends: 'There is a most splendid paradox in Art that often the wildest, most far out, random unprogrammed activity can in the end produce work which may exercise the most profound and fruitful control of the human situation.'[49] A controlled systems pedagogy then, was simply a beginning – the result was dependent on the participants. Just as early cybernetic machines were messy, unpredictable and often surprising, so are any group of individuals undertaking a controlled exercise. This chapter concludes with three Groundcourse projects and events which demonstrate the behaviourism that informed the course.

A ripple in the aether: control tactics in Groundcourse

In 1963, the Groundcourse student John Bonehill, together with two of his classmates, staged a fake bombing in the London Underground. The fake bomb was set up in Mansion House station, during a period of heightened military tension as the Cuban Missile Crisis inched towards its zenith. In the brief furore that followed, the students were suspended from Ealing College for two weeks although they escaped any criminal charges. This reckless criminal act was their response to a set of project directives that asked groups of three students to:

> Go out into town, & cause a ripple in the 'ether', cause disruption by any means available, i.e. cause a traffic accident, or commit robbery, anything which will cause a reaction.[50]

Looking objectively at the project description that Bonehill recalls, it is easy to see how the group derived from these instructions the impetus to create genuine shock and unrest. The examples – 'cause a traffic accident' or 'commit robbery' – did not ask for performance, but for violent interventions into city life. To cause a 'ripple' then, was to introduce an element of chaos into a system, a beat of the butterfly's wings, that would ripple outwards causing change. From a contemporary perspective, it seems both dangerous and foolish to set such a task for foundation students. Today the repercussions for such an action would have at the least been a prison sentence, with a real threat to the personal safety of the protagonists. This model of art as direct intervention took shock tactics to a new level; rather than shocking the students through a series of formal disruptions to the material processes of abstraction, the students themselves became agents in an act of social disturbance.

This exercise reflected a much bigger trend in London towards destructive forms of politicised art. In 1962 Gustav Metzger gave a lecture for staff and students of Groundcourse Ealing, during which he showed '[…] a succession of about fifty slides representing art, society, space research and war taken from newspapers and magazines, as well as two films on automatic mechanical self-replication and the movement of cells', in the words of Andrew Wilson.[51] At this event, Metzger discussed a number of works of contemporary art in relation to the threat of nuclear annihilation, sifting through media images and drawing destruction, creativity and war into a layered and bewildering embrace. From Wilson's description, it is apparent that Metzger's presentation brought in both the biological and the mechanical in a way that was appropriate to the interests of the audience. Several staff and students were themselves interested in Metzger's work and the 'ripple in the aether' project certainly had an undercurrent of auto-destructive intent. Considering Metzger's explicit address of the nuclear threat, it is evident that both Groundcourse staff and students understood the fact that methods of social disruption were innately political; they were shock tactics that played upon, emphasized or subverted the tense order of public life. In this period, Metzger's manifestos were densely layered with jarring ideas and contemporary references that held together with a tense energy. The words feel collaged, connected to the same impulse that underwrote the collage work produced by members of the Independent Group such as Paolozzi's *BUNK!*, McHale's *Machine Made America* or Henderson's *Head of a Man*. In his Manifesto Auto-Destructive Art he writes:

> Man in Regent Street is auto-destructive.
>
> Rockets, nuclear weapons, are auto-destructive.
>
> Auto-destructive art.
>
> The drop drop dropping of HH bombs
>
> Not interested in ruins, (the picturesque).
>
> Auto-destructive art re-enacts the obsession with destruction, the pummelling to which individuals and masses are subjected.
>
> Auto-destructive art demonstrates man's power to accelerate disintegrative processes of nature and to order them.[52]

It was a manifesto of the moment; a statement of compressed layers about the end of modernity, the nuclear threat and the emergence of a new art so subversive that it was self-destructive. In Metzger's view, the great machine of art in modernity had destroyed the past as it created in the present; auto-destructive art did both simultaneously, an uncanny coupling of creation and destruction that left only detritus. Metzger described a planned ruin, as perfect a destruction as 'the drop of HH bombs'.[53] It was an art that 'mirrors the compulsive perfectionism of arms manufacture—polishing to destruction point'.[54]

Metzger was a member of the British anti-war group the Committee of 100, an organisation that aimed to create publicity for the anti-nuclear campaign through the kind of protests and acts of civil disobedience that would attract press attention. The protests that this group arranged were essentially performative – non-violent mass demonstrations that attracted press attention and caused disruption. Its public signatories included Bertrand Russell, Ralph Schoenman and Reverend Michael Scott. Russell wrote in 1961 that:

> To make known the facts which show that the life of every inhabitant of Britain, old and young, man, woman and child, is at every moment

> in imminent danger and that this danger is caused by what is misnamed defence and immensely aggravated by every measure which governments pretend will diminish it – to make this known has seemed to some of us an imperative duty which we must pursue with whatever means are at our command.[55]

He described 'kernels' of American personnel at British Base Camps, kept entirely separate from their peers and trained for fast response to disaster, receiving their orders direct from Washington. He discusses the opinions of a panel of defence experts at the Ohio State University and the proceedings of the Pugwash conferences, demonstrating a prevailing belief that nuclear war was all but inevitable. He quoted Sir Charles Snow, who had recently written in *The Times* that 'Speaking as responsibly as I can, within, at the most, ten years from now, some of these bombs are going off. That is a certainty.'[56] In what amounts to a call to disarm, Russell writes:

> If all those who disapprove of government policy were to join in massive demonstrations of civil disobedience, they could render governmental folly impossible and compel the so-called statesmen to acquiesce in measures that would make human survival possible.

The organization was behind a second small activist group called 'Spies for Peace', who in 1963 broke into the RSG-6 bunker in Reading, obtaining, copying and publishing documents proving the existence of the network of fourteen regional bunkers.[57] The Spies for Peace were never arrested, although there were unproven suspicions about their identities, and it wasn't until the twenty-first century began that four individuals identified themselves.[58] When the Spies for Peace managed to break into RSG-6 undetected, they immediately published the pamphlet *DANGER!! Official Secret. RSG-6* in which they detailed their discoveries. There was a public outcry that located itself around two poles – firstly, there was outrage that they could have risked national security to this extent. Secondly, there was disbelief, anger and disappointment that in the event of war, the Government would abandon its people and retreat below the ground. It reads:

This pamphlet is about a small group of people who have accepted thermonuclear war as a probability, and are consciously and carefully planning for it.

> They are above the Army, the Police, the Ministries or Civil Defence. They are based in fourteen area headquarters, each ruled by a Regional Commissioner with absolute power over millions of people. In the whole of Britain only about 5000 men and women are involved: these chosen few are our shadow military government.[59]

This strain of activism undertaken by the Spies for Peace sought to uncover this new, intensely secret mode of military operation and in doing so force the public to confront the real possibilities and potential consequences of nuclear war. Considering the covert activities of Spies for Peace alongside the public and performative protests organized by the Committee of 100, it is evident that the politics of the moment revolved around the tension between state secrecy and public activism. Metzger's auto-destructive art was closely aligned with his anti-nuclear stance, operating somewhere between hope and despair, nihilism and activism. The close embrace of post-war art and Cold War politics has had curiously scant academic attention, perhaps because Trauma Theory provides such a seductive framework for 'before' and 'after' World War II, creating an artificial break in history.

The Cold War then, was both secretive and pervasive in its influence upon British culture. This was certainly true of art pedagogy; the overlapping tendencies of controlled exercises, surveillance, behaviourism and game-play bear an indelible imprint of Cold War politics. It must be noted that auto-destructive art was verging upon constituting a movement in London at the time; behind the framing concept, Metzger's manifesto held together the politics and the fears of the age. A few years later, Ascott himself was an invited attendee of the 'Destruction in Art' symposium of 1966, open only to those artists who had 'used *actual destruction of materials* as part of their technique'.[60]

Bonehill's fake bomb was a hoax, but even as such it left nothing behind but disruption: delays in the public transport system, public anxiety and demands upon the police service. The exercise had explicitly sought disruption, creating a framework of practice for the young students that positioned them as protagonists in a form of social protest. However, the exercise did not stop there. The students had to 'come back and then write an essay describing what happened. Make a drawing and a diagram then build a machine that expresses the essence of the incident.'[61] In this exercise then, the students were first active and performative, then analytical. The instructions to draw, diagrammatize, write and build an account of the incident direct the students to explore different means of communication, from written language to mechanisms. It should be noted that the students were asked to build machines, not sculpture – a machine must consist of different parts moving together to create a given function. This is an interesting method of analyzing the kind of combustive energy that drove the first part of the task. The 'machine' as an outcome of this mode of performance or social intervention was evidently intended to make students consider their project as a system. Groundcourse pedagogy directed the students towards a more connected understanding of both making art and viewing art as processes, interlinked and potentially endless.

Townshend recounts attending college on the day, in 1963, when the Cuban Missile Crisis came to a head. He recalls students crying, laughing hysterically, or sitting in silence. He himself chose silence and after what felt like a brush with death, chose to seize the day and reignite his dalliance with Carol Daltrey, albeit briefly.[62] Eno was still at school that day; students were called into assembly and Eno remembers a pervasive and heavy certainty that the world was ending.[63] In that very real moment of fear, we might consider the background pressure of the Cold War and its significance for that generation of students. There was a divide between those that taught and those who were students – teaching staff had, with very few exceptions, either completed their national service or served during World War II. By the start of the 1960s, the students, with very few exceptions, had not. As we unpick these behaviourist art models, the place of military service within their post-war evolution is surely a significant one. Control and observation as teaching mechanisms relate visually, dynamically and practically to military operations and training, while the overtly emotional and performative student responses to these exercises speak of their dislocation from this generation above them. Given that within a few years of Groundcourse there would be significant student unrest in the 1968 protests, we must place behaviourist practices within this generational debate about personal creative control, ownership and, eventually, freedom.

An interesting aspect of these exercises in confined creativity is the fact that teaching staff had no idea how students would react and it was as much a learning experience for them as for the students. While the relative allowances of freedom and control are the defining elements of any pedagogy, it was not the intention of either experiment to simply limit or control what students learned. Rather, it was to provoke unease and to explore the tension of observation. While the pedagogical underpinning for exercises such as these was to engage students with their own reactions to limitations and possibilities within their environment, the exercise was created in the spirit of mutual experimentation. While both student and teacher had little idea of what the outcomes might be, there remains an issue of relative power and agency for those involved. While these exercises in self-analysis were intended to make students analytical and thoughtful in their responses, the process of scrutinising personal patterns of behaviour can be challenging, intimate and painful. Behaviourism was innately detached; Skinner's lab rats were scrutinized as living models for human cognitive processes. This detachment characterized the application of behaviourist science to human subjects; controlled circumstances of some kind were applied in order to condition the responses of the subject. Skinner's concept of operant conditioning advocated modification by reinforcement and punishment. Whether intended or not, in a formal behavioural sense, Eno's spell in a trolley constituted punishment rather than reinforcement; it was restraint. We might also interpret this process of behaviour modification as a militarized form of group control. The concept of creating adaptive systems of behaviour in order to benefit group function has much in common with the selfless operation of military units, in which individuals are complicit in their own abdication of personal power.

The relationship between cybernetics and behaviourism is certainly apparent here and it certainly has echoes of World War II, during which the face of military service changed, with men and women part of larger systems of operation, both living and technological. If we view the post-war heyday of behaviourism through this lens, it becomes clear that the science of behaviour was about maintaining power of a collective, for military, economic or social benefit. The individual was unimportant to the point of insignificance in this model of psychological control; in fact, the behaviourist approach was to objectively control – or utilize – individual behaviour for larger benefit. Like Ascott's concept of collective as 'organism', the individual was part of a system of operation with a larger task in mind. In the context of art education, which has largely managed to maintain its quasi-mythical objective of individual creative practice, this spell of collective action is all the more notable. It was steeped in the Cold War tensions of control and surveillance, something that becomes increasingly clear as we reflect on these episodes of art-surveillance.

Ascott had read Skinner's 1953 book *Science and Human Behaviour*, in which he had outlined the ways in which human behaviour patterns were established in relation to their environment:

> Certain processes, which the human organism shares with other species, alter behaviour so that it achieves a safer and more useful interchange with a particular environment. When appropriate behaviour has been established, its consequences work through similar processes to keep it in force. If by chance this environment changes, old forms of behaviour disappear, while new consequences build new forms.[64]

This idea was the core of his book, which explored the potential for developing a science of behaviour, particularly with reference to the kind of Pavlovian behaviourist experiments outlined at the start of this chapter. Skinner gave an analysis of behaviourist technique, from the individual to the collective and within the parameters of existing models of social control such as religion, government and economics. He described the organism in what appear retrospectively to be distinctly cybernetic terms: simultaneously influenced by, and influencing, its environment. This feedback relationship was based on Skinner's observations of reflexes as 'a correlation of stimulus and response', an idea that would later be taken to a new level in the form of Maturana and Valera's autopoietic – or self-reflexive – systems.[65]

On the fertile interdisciplinary breeding ground of cybernetics, biology and sociology, behaviourism treated systems of control as a biological problem, human life reduced to organism within a system. Skinner observed that '[…] machines have become more lifelike, and living organisms have been found to be more like machines. Contemporary machines are not only more complex, they are deliberately designed to operate in ways which resemble human behavior.'[66] This mode of thought had a trajectory from the interwar years. Tracing a line from the mechanized biology of D'Arcy Wentworth Thompson to post-war computers which could perform tasks previously only achievable within the human brain, there was a moment of exchange in early cybernetics between organism-as-machine and machine-as-organism, a technologised world-view in which systems of behaviour could be detachedly observed and controlled.

When Skinner had published his first book on the subject, *The Behaviour of Organisms*, in 1938, he introduced the concept of respondent and operant behaviour. The focus on science in this later book can be read in the wider context of post-war interactive technologies. Given the paranoia of the Cold War environment, a science of predicting and controlling human behaviour was an attractive prospect. The dark potential of behaviourism was explored in popular culture too – a notable example being the 1959 novel *The Manchurian Candidate*. The difficulty of behaviourist science arises from the fact that somebody – a person or agency – must always manipulate the responses and resulting behaviours of the subject. Behaviourism operated upon a premise of objective control.

Power relationships in pedagogy: the quadrangle incident

This principle of objective control came to life in Groundcourse teaching experiments, which saw staff scrutinize the actions and reactions of students in certain created conditions. In *Re-Make/Re-Model: Becoming Roxy Music*, Michael Bracewell included interview material with Brian Eno pertaining to his time as a student of Groundcourse at Ipswich.[67] Eno described the level of psychological manipulation which underpinned the course, offering several anecdotes about experiments and classes. He described an incident in which there was a notice instructing students to assemble in the quadrangle at 9.30. They did so and suddenly all the doors were locked behind them. The staff then assembled with chairs on the flat roof and studied them. Eno recalls that as the students began to comprehend their circumstances, a spectrum of extreme responses emerged, including destruction, anger and upset. In the spirit of collective behaviour, Eno himself set up a committee with other students.[68]

It is worth noting that this experiment predated Peter Kardia's 'locked room' pedagogical project at St Martin's by at least four years.[69] Kardia's experiment was extensive, lasting a full term and students voluntarily turned up in the morning only to be locked in the white space of their studio for the morning. The same precept of confinement ties Kardia's experiment to the quadrangle experiment at Ipswich: each tested how students would react to their liberty being restricted and their actions scrutinized. Beyond this, Groundcourse in its entirety operated upon a model of contained experiment; students were informed from the outset that the course was a process, that they were part of a larger entity for the two-year duration of their studies. Eno recalls:

> Then Tom Phillips read this text out, I think over a megaphone, and it was a quote from Lenin; it said something along the lines of 'You are

> worse than chickens. A chicken will be trapped inside a chalk circle, but you have drawn your own chalk circle and trapped yourselves.' And that was the only thing they said.[70]

This behaviourist message conveyed the idea that the unquestioning acceptance of instruction and authority had the potential to be a trap. The students had seen a notice that bore instructions, and they had assembled as instructed. The doors had closed and they were trapped by their obedience. Of course, this was just the beginning of the exercise – trapping the students was not a way of making a point, but a point of departure for them. Skinner had written: '[...] If a cat is placed in a box from which it can escape only by unlatching a door, it will exhibit many different kinds of behaviour, some of which may be effective in opening the door.'[71] Similarly, the experience provoked many kinds of behaviour from the students, ranging from destructive, emotional or rational approaches to their predicament.

Tom Phillips was the tutor of a liberal arts course for Groundcourse students at Ipswich, where he formed a close friendship with his student Brian Eno. The chalk line notion appears to have derived from Vladimir Lenin's famous train journey from Switzerland to Russia in 1917.[72] During the journey, Lenin had insisted that a chalk line be drawn inside the train between the Bolsheviks and the German Officers accompanying them on their journey, creating an artificial and symbolic boundary between two hostile territories as they travelled, together, across Europe. Phillip's chalk line holds dual connotations of the limits we impose upon ourselves through patterns of behaviour, and, simultaneously, the implications this holds for art production. The chalk line was also a statement of power, of division and of outside control. It was designed to make the students question their own agency; one might simply step over the line and render it meaningless. However, Lenin's chalk line was also a physical display of a division that already existed, palpable and concrete to both the Bolsheviks and the Germans. It was an exercise designed to make students think critically about the power relationships between staff and students, about what limited their own decision-making processes and about free thought.

Art training is unlike any other form of education, in that the end goal is creative self-sufficiency and the ability to generate new work entirely independently. The students were to be creators of their own style, their own techniques and eventually, their own oeuvre. While this ideological model of art education persists to this day, even during the process of academicisation that the subject underwent in the 1960s, this vision of solitary practice was something few would fulfil. Many of the artists that entered into the new diploma courses went on to supply the need for teaching staff for the next cohort of DipAD students as further colleges gained approval to deliver the course. Behind Phillip's chalk line event, we glimpse the fractured past and unknown future of art education, as he questioned the teacher–pupil relationship and its relevance in training contemporary artists in an age of rapid change. While Basic Design tutors had ripped up student drawings in order to shock them into rethinking process, in Groundcourse it was the self that was to be reconstituted through phased processes of self-analysis.

Product or outcome mattered little within the very particular confines of Groundcourse, where completion meant passing and grading was an irrelevance. The scores of student drawings, sculptures, games and other objects were not to be measured for their own worth or against each other; they were by-products of a process, and the process was the achievement. Throughout the course, it was the students themselves that received the close scrutiny that would have previously been reserved for their work, a measure of the instability of values in the visual arts and the related and inevitable conceptualization of art education. The first tranche of students to take the new route through foundation studies to DipAD did so at a time when values were in flux. The chalk line is a suitable analogy for a mode of controlled education that was in danger of being erased.

The locked quadrangle at Ipswich and its underlying play upon the politics of personal freedom provoked the students into a self-conscious, conceptual mode of art practice. Eno described the range of behaviour this act of imprisonment inspired from the trapped students, from shouting and threatening to smash windows to group protests and banging. They could have sat quietly and waited for the three hours for which they were held captive, but the tension of captivity coupled with the overhead surveillance by their teachers provoked these more self-conscious and performative reactions. Aside from the individual learning processes that were taking place, we must address the structure of this activity and its larger implications. It was an exercise in relative power, and powerlessness. As a form of conditioning, there are a number of levels upon which it operates. Firstly, there is captivity and the withdrawal of personal freedom. While the students involved fully understood that they had been captured for the sake of performance or reflection, they also immediately recognized that the premise they had been given was captivity and surveillance. Their responses were conditioned by this knowledge. This was tied to the hierarchical structure of the exercise; just as the tutor parades the art studio, observing and judging, the teaching staff sat in leisurely poses, observing through binoculars. It became a performance of a power relationship.

Eno recalls that in a counter-exercise that followed the Quadrangle Lock-In, students were invited to watch as their tutors painted live in front of them and to heckle, comment or deride as they saw fit. The commentary verged on the aggressive, aping the notoriously hard lines of the 'studio crit' of the day.[73] This retaliatory exercise presumably allowed students to switch roles and hence consider the frameworks of convention within which a studio education operated and to consider what alternatives there might be. Ascott had already forged Groundcourse on an alternative, in that all students who completed the course would pass and there would be no variants to the level of pass. These performances of criticism therefore served as an investigation into the principle of value-judgement, both conceptual and aesthetic, which informs the education model within the school of art.

Marbles on the floor: disorientating tactics and conditioning

One night in 1962, the students and staff staged a series of performances in the theatre of Ealing College of Art. In between each performance, the large audience was plunged into pitch darkness for indeterminate spells, before the stage lights came on dazzlingly and the show continued. After the last performance ended, the darkness returned and the audience sat waiting, first patiently and then with growing impatience and annoyance, for another performance or an announcement. None came. Eventually the audience stumbled and staggered to the door, bursting out and emptying into the hall. However, the floor of the hall was scattered with marbles. Half blinded by the transitions from dark to light and back again, disorientated and confused, people staggered, slid and fell. Ascott left by taxi and passed a friend in another cab who rolled down the window to ask what on earth was happening at Ealing, having seen people lurching and tripping down the stairs.[74]

This performance-experiment used light and darkness to disorientate its unwitting subjects, veering from the sensory overload of the performances to the enclosing stillness of the darkness. The effect of light pulses was in itself the subject of cybernetic research; Walter had investigated how strobe lighting had a mind-altering effect, including the induction of epileptic seizures.[75] Walter later noted that '[...] Illusory experiences produced by flashing lights' were used in subcultural contexts as a '[...] standard method of stimulation'. He joked that '[...] I should be paid a royalty because I was the first to describe these effects'.[76] Flicker and strobe effects were, as Pickering has described, used in a number of countercultural contexts, often in conjunction with hallucinogens. In this sense, light effects formed part of a convergence of psychiatry, cybernetics and psychedelia that informed British counterculture. This theatre performance and its stumbling, painful aftermath were certainly based upon the same principle of using light to alter sensory responses – also in this case creating physical incapacity.

This experimental performance drew in the audience, blurring the distinction between actor and viewer and making both into participants. It was certainly informed by behaviourist ideas; in this case the audience were the subject of behaviour modifying influences. The living creature subjected to behaviourist control is influenced by means that affect its immediate environment and its ability to survive within that environment, such as food or free movement. Light, sound or sensation are means of behaviour modification – they serve as signals that reinforce 'good' behaviour or disrupt 'bad' behaviour. In this sense, they are code. The sequence of application becomes pattern and the operant learns the pattern. At this performance event, the audience's expectation that they were to watch a spectacle was actively disrupted using light and dark, causing confusion that was both physical and cerebral. As the audience stumbled and fell in a hall of marbles, it was evident that the cerebral disturbance had been successful – perhaps too successful. Fortunately, there were no serious injuries that night. However, it should also be noted that this was not an attempt to control behaviour, but rather to disrupt it. The performance drew in contemporary influences such as John Cage, using darkness in the manner that Cage had used silence in 4'33", in that the blackouts created periods of awkward self-awareness that disrupted the comfortable anonymity of the group audience. As these early pedagogical experiments in control evidence then, the most established programmes of behaviour modification might be subverted, ignored or undone. The living are creators and destroyers of systems of control. Ascott's student 'organisms' operated on this very premise – we are possibility. As long as we recognize the patterns that make up our *selves*, we might subvert this pattern and in doing so, send ripples out into the aether. When John Bonehill staged his fake bombing, he exposed not only the potential for a system of education to be subverted, but also the growing tension that the Cold War was provoking in a generation of young men and women.

Notes

1 Kate Sloan *Interview with Roy Ascott* (September 2012).
2 *Ibid.*
3 The two 'Coldstream Reports' of 1960 and 1970 marked a period of policy changes in art education. See William Coldstream (chair) (1960) 'First Report of the National Advisory Council for Art Education' and (1970) 'Joint Report of the National Advisory Committee on Art Education and the National Council for Diplomas in Art and Design'. For an overview of the history of art education policy in Britain, see Clive Ashwin (1975) *Art Education: Documents and Policies 1768–1975*. London: SRHE.
4 Lynda Morris (2013) 'The Impersonal Eye of the Camera'. *Bare Life: Bacon, Freud, Hockney & Others. London Artists Working from Life 1950–1980 (Exhibition Catalogue)* LWL-Museum für Kunst und Kultur, Munster. Munich: Hirmer Verlag.
5 Coldstream (1960) 96.
6 Rose Rose, former girlfriend of Freddy Mercury (2007) Email exchange with the author.
7 Conscription in the United Kingdom during and following World War II ran from 1939 until 1960.
8 Anton Ehrenzweig. Unpublished draft typescript. William Johnstone Archive, National Library of Scotland Dep. 332/6.
9 'Late Middle English: via Latin from Greek paidagōgos, denoting a slave who accompanied a child to school (from pais, paid- 'boy' + agōgos 'guide')' From Oxford Dictionaries online, accessed 12/02/2017: http://oxforddictionaries.com/definition/pedagogue.
10 This definition of 'education' is from the Oxford English Dictionary.
11 Association of Staff and Students of Hornsey College of Art (1968) *The Hornsey Affair*. London: Penguin.
12 Kate Sloan (2012).
13 See Andrew Pickering (2011) 11–12 for an account of Beer's influence upon Brian Eno.

14 *Ibid.*
15 Ashby (1960) and Beer (1970).
16 J.H. Clark (1963) 'Adaptive Machines in Psychiatry'. *Progress in Brain Research*, Vol. 2, 224–35.
17 *Ibid.*, figures 7 & 8.
18 *Ibid.*, 228.
19 B.F Skinner (1963) 'Behaviourism at Fifty'. Reproduced in: A. Charles Catania & Stevan Harnad (eds) (1988) *The Selection of Behaviour: The Operant Behaviourism of B. F. Skinner: Comments and Consequences*. Cambridge: Cambridge University Press, 344–5.
20 Letter from Anthony Blake to Roy Ascott. October 14, 1964. Collection of the artist. Unpaginated.
21 He has since written several books on systems thinking and mysticism as well as founded the journal *Systematics.*
22 *Ibid.*
23 *Ibid.*
24 Stafford Beer (1959) *Cybernetics and Management*. London: English Universities Press, 125. Also quoted in: Andrew Pickering (2010) *The Cybernetic Brain: Sketches of Another Future*. Chicago: University of Chicago Press, 327.
25 For an account of the project, see Cedric Price & Joan Littlewood (1968) 'The Fun Palace', *The Drama Review: TDR*, Vol. 12. No. 3, 127–34.
26 Letter to Roy Ascott from J. Rose. June, 1968. Invitation to submit a paper for the Institute of Computer Sciences' *International Conference of Cybernetics.* Archival collection of Roy Ascott.
27 Letter to Roy Ascott from J.G. Lewis, 14 October 1969 and Letter to Roy Ascott from the Birmingham College of Art and Design, 19 March 1969, both archival collection of Roy Ascott.
28 Roy Ascott. *Provisional Planning Document for Ipswich Symposium.* Archival collection of Roy Ascott.
29 *Ibid.*
30 International Congress of Cybernetics. *Letter to Members ahead of I.C.C London*. February, 1970. Archival collection of Roy Ascott.
31 Stephen Willats in: Bronac Ferran (2017) 'The Conceptual Design in 1965: Stephen Willats Interviewed by Bronac Ferran'. *Interdisciplinary Science Reviews*, Vol. 42. No. 1–2, 201–13, 212.
32 Stephen Willats (1965) 'Editorial'. *Control Magazine*, Issue One, 1965.
33 Stephen Willats, quoted in: Gwen Allen (2011) *Artists' Magazines: An Alternative Space for Art*. Cambridge: MIT Press, 251.
34 Ascott commented upon how problematic this situation is in: Roy Ascott and Kate Sloan *In Conversation*, 8 March 2017, The Henry Moore Institute, Leeds.
35 Stephen Willats, quoted in: Ferran (2017) 'The Conceptual Design in 1965: Stephen Willats Interviewed by Bronac Ferran'. *Interdisciplinary Science Reviews*, Vol. 42. No. 1–2, 201–13, 201.
36 Faure Walker, quoted in: John A. Walker (2001) *Left Shift: Radical Art in 1970s Britain*. London: I.B. Tauris, 101.
37 *Ibid.*, 101.
38 Walker (2001) 101.
39 The National Physical Laboratory hosted residencies for several British artists, including Stephen Willats, with a similar ethic of promotion and dissemination to the Bell Laboratory residencies in the US.
40 Matthew Wisnioski (2013) 'Why MIT Institutionalized the Avant-Garde: Negotiating Aesthetic Virtue in the Postwar Defense Institute'. *Configurations*, Vol. 21. No. 1 (Winter) 85–116, 90.
41 *Ibid.*, 112.

42 Roy Ascott (1965) 'Statement'. *Control Magazine*, Issue One. 1965.
43 *Ibid*.
44 *Ibid*.
45 Pickering (2010) 25–6.
46 See Paulo Freire (1970) and Henri Giroux, (1983) & (1991).
47 Henri Giroux (1991) 'Border Pedagogy and the Politics of Modernism/Postmodernism'. *Journal of Architectural Education*. Vol. 44, No. 2, 69–79.
48 *Ibid*., 72.
49 Roy Ascott (1965) 'Statement'. *Control Magazine*, Issue One.
50 John Bonehill (2011) Email exchange with the author.
51 Andrew Wilson (2008) 'Gustav Metzger's Auto-Destructive/Auto-Creative Art: An Art of Manifesto, 1959–1969'. *Third Text*, Vol. 22, Issue 2, 177–94, 190.
52 Gustav Metzger [10 March 1960] 'Manifesto Auto-Destructive Art' reproduced in Kristine Stiles & Peter Salz (eds) (2012) *Theories and Documents of Contemporary Art* Berkeley: University of California Press, 470–71.
53 *Ibid*., 470.
54 *Ibid*., 471.
55 Bertrand Russell (17 February 1961) 'Civil Disobedience'. *New Statesman*, 17.
56 Charles P. Snow (1960) 'The Moral Un-Neutrality of Science'. *Public Affairs*, 1971 [1960]. London: Macmillan, 187.
57 R. Jeffreys-Jones (2013) *In Spies we Trust: The Story of Western Intelligence*. Oxford: Oxford University Press, 156.
58 See: Natasha Walker (May 20, 2002) Essay: 'How My Father Spied for Peace'. *New Statesman*.; Mike Hastings (21 March 2010) 'Nuclear Bunker Spy Comes out of Hiding'. *The Times*; and Sam Carroll (2010) 'Danger Official Secrets and the Spies for Peace: Discretion and Disclosure in the Committee of 100'. *History Workshop Journal*, Spring (69), 158–76.
59 Spies for Peace (1963) *DANGER! Official Secret RSG-6*.
60 Invitation to the 'Destruction in Art Symposium'. Roy Ascott Archival collection.
61 John Bonehill (2011) Email exchange with the author.
62 Townshend (2012) 50.
63 Kate Sloan *Interview with Brian Eno* (Edinburgh, May 2016).
64 B.F. Skinner (1953) *Science and Human Behaviour*. New York: Macmillan. 1.
65 Humberto R. Maturana & Francisco J. Varela (1980) [1972] *Autopoiesis and Cognition: The Realization of the Living*. Dordrecht: D. Reidel Publishing Company.
66 *Ibid*. (1953) 49.
67 Michael Bracewell (2007) *Remake/Remodel: Becoming Roxy Music*. Cambridge: De Capo Press, 195–208.
68 Kate Sloan *Interview with Brian Eno* (Edinburgh, May 2016).
69 See Hester Westley (2007) 'The Year of the Locked Room'. *Tate Papers*. Issue 9. Spring 2007. Accessed online 12 Oct 2015: http://www.tate.org.uk/context-comment/articles/year-locked-room.
70 Bracewell (2007) 205.
71 Skinner (1953) 49.
72 Winfried Wolf (1996) *Car Mania: A Critical History of Transport*. London: Pluto Press, 62.
73 Kate Sloan *Interview with Brian Eno* (Edinburgh, May 2016).
74 Roy Ascott and Kate Sloan *In Conversation* 8 March 2017. Henry Moore Institute, Leeds.
75 For a full account, see Andrew Pickering (2007) 77–82.
76 *Ibid*., 77. Quoting Geiger (2003) 83.

5 Calibrator

Self-consciousness | adjustment | possibility

Shock

Several years after participating in Groundcourse, the British musician Pete Townshend stabbed his hand with a steak knife, drawing a line in blood on the wall. 'THAT'S a line!' he shouted, to much applause.[1] In the blur of a rock and roll party, the hotel's insipid choice of bedroom art had incited criticism by those present, prompting Townshend to enact a moment from early in his own art training that had stayed with him over the years. He had been a student of the first incarnation of Groundcourse at Ealing College of Art, amongst the first cohort of UK students who had been required to complete a foundation course before proceeding to study for a diploma in art or design. In the first week of the course, two sessions on draftsmanship saw Townshend and his peers exposed to radically conflicting concepts of drawing practice. The first tutor gave the class precise instructions for a first line that had to be 'north to south, six inches long, of uniform thickness and drawn with a 3B pencil without a rule'.[2] Townshend recounts that a second lesson brought him into contact with Anthony Benjamin, who asked the class again to draw a line. When he saw the identical results produced by the students, he left the room and returned with his colleague, Brian Wall. The pair shouted and raved at the class before:

> [...] Benjamin produced a small penknife and pricked his finger, dragging blood across a white sheet of paper. That's a line. DO you understand? Of course we understood. We were the innocent victims of a struggle between the old and the new.[3]

In Townshend's recollection and subsequent bloody re-enactment, this moment in his education was charged with drama, underwritten by the profound sense that change was afoot. The deep-rooted tension that has always existed between tradition and progress in the school of art was certainly more marked in the post-war years. In an era of extensive policy development, the study of art and design was changing rapidly and at the same time, many young staff were hired to meet the curriculum requirements at foundation and DipAD level. This need for new staff certainly created opportunities for newly qualified artists such as Ascott. The foundation courses in particular became zones of experimentation for young staff – unlike the diploma course, there were no formal guidelines for foundation courses. Basic Design and associated pedagogical approaches had dominated art teaching in the 1950s. The foundation course had been introduced due to the perceived success of these 'basic courses' in order to bridge the ever-widening gap between a conventional secondary art education and the art school after abstraction.[4]

As a generation of artists graduated and took up posts in foundation courses across the country, it quickly became evident that a number of these courses had departed from the conventions of abstract art education and were increasingly experimental, performative and conceptual. Ascott hired for what he termed 'requisite variety', on the understanding that the different approaches and beliefs of his staff would produce a more dynamic and effective system.[5] In Benjamin and Wall's battle cry in the war over line, the abandonment of the precisely drawn draftsman's line for a line dragged in blood represented a move from objective, material construction to an embodied, performative physicality of process – a blurred and transitory space between media-led education and the primacy of what the young Ascott thought of as *ideas*.[6] As Ascott and many of his contemporaries moved towards conceptual practices, it quite naturally caused reverberations within schools of art.

Like Basic Design, Groundcourse set out to 'undo' the presumptions about art making that students had gained within the more conventional and figurative confines of secondary art education. John Bonehill, a former Groundcourse student at Ealing College, recalled one of his first classes at the college, taught by Denis Bowen:

> He [...] said 'from now on you will forget everything you ever learned about art, no more blue skies and green trees for you'. We each were given a block of black wax and three sheets of brown paper and we were to spend the first day 'Brass Rubbing' our way around campus, getting rid of our preconceptions.[7]

This exercise immediately did away with the detachment and distance of drawing as taught in secondary school, where the majority of students had experienced drawing as a disciplined observation exercise. Instead, Bowen sent students to gather imprints directly from form, doing away with observation and replacing it with an exercise in process. As directed by Ascott, the first years were to be challenged by tutors through exercises and activities designed to make them reassess their preconceived notions of visual art practice. Bonehill himself turned up for the first day of Groundcourse in his school uniform – a fact that is perhaps a good measure of the leap to be taken from a 1960s secondary comprehensive school to a radical art training. He was, in his own words, 'sixteen years old and still an active member of the boy scouts'.[8] Bonehill recalls attending one three-hour class on '[...] how to correctly sharpen a pencil – very zen'.[9] His recollections of Groundcourse rotate around interconnected moments of shock as well as the personal liberation prompted by the subsequent inventions, processes and experiments. It was a process designed to gradually erase one mode of thinking without immediately providing another. While the tumultuous dance of rejection and suggestion that had characterized modern art so often relied on the disavowal of existing values or ideologies, there was a point of difference in Groundcourse; Ascott wanted to maintain uncertainty, not to replace it. Ascott wanted his students to be self-aware, questioning and open. He wanted them to be aware that every situation was built upon decisions; that in essence, every situation was possibility.

Reprogramming students: systems in drawing exercises

In 1964, Ascott published the essay 'The Construction of Change', in which he explained his cybernetic theory of art before outlining the Groundcourse curriculum with descriptions of various exercises.[10] In the first year Groundcourse curriculum, a number of drawing exercises directly explored systems of biological and technological interaction. With echoes of the Basic Design shock tactic of tearing up student drawings, one exercise prompted students to:

> 1. Draw a man, machine or animal. Cut up the drawing into seven sections (e.g. arm, head, wheel, handle, etc.). Put the pieces with everyone else's in a box. Pull out another seven at random; logically construct a new entity. Draw the environment in which you might expect to encounter it.[11]

Ascott describes this as an exercise in 'behaviour, environment and identity'.[12] The resulting drawings would have been a series of collaged cyborgs not unlike the assemblage heads and bodies in the manner of Paolozzi that were created by students of Richard Hamilton at King's College.[13] However, Ascott took the theme further by introducing an element of chance and then leading the students towards considering, once again, the way something might live in a given environment. This process focused the students upon how the form of both living and engineered systems was analogous to their function and in turn, upon how form was suited to environment. The exercise itself was also a system, in which other students contributed to each individual outcome. This exercise dissolved boundaries between living and man-made mechanisms, a proposition that was deeply rooted in the interdisciplinarity of post-war British art. In many of Richard Hamilton's exhibition projects, including *Growth and Form* and *Man, Machine and Motion*, as well as other Independent Group exhibitions such as Paolozzi and Henderson's *Parallel of Life and Art*, the curators employed suggestive visual comparisons between natural and engineered phenomena, suggesting, rather than defining, the similarities between natural and manmade forms and processes.[14] Ascott was, however, the first artist in Britain to explicitly articulate a systems approach, in word, art and in pedagogy, and to do so in direct reference to cybernetic theory. Ascott's cybernetic model for art and pedagogy placed him at the forefront of the worldwide field of cybernetic art, articulating his ideas in theoretical terms at a time when most artists were simply experimenting with the potential of new communication technologies.

This integrated systems approach led to linked drawing exercises that often referenced that recurrent concern of first wave cybernetics – the relationship of the organism to its environment, which Ashby had described as '[...] partly conceptual, and to that extent arbitrary', creating a situation in which 'Variables within the body may be justifiably regarded as the 'environment of some other part'.[15] Ashby's words evoke a complex world of interacting systems, opening endlessly outwards, and Ascott's drawing exercises often played upon the relationship of organism to environment. In another exercise, students were told to 'Invent a typewriter-bird and show the kind of tree within which it could most successfully hide'.[16] While there is a transparent element of fun here, the underlying message is familiar – living things and machines adapt to their environment and are suited to it. To understand any organism or any machine, one must also understand its environment and in considering this relationship, the world opens out into converging systems. This message was repeated throughout the first-year curriculum, and emphasized further by exercises which focused on systems awareness too. On the subject of designing – or programming – adaptive systems based on the human brain, Ashby had written that:

> [...] a system can be both mechanistic in nature and yet produce behaviour that is adaptive. I hope to show that the essential difference between the brain and any machine made yet is that the brain makes extensive use of a method hitherto little used in machines. I hope to show that by the use of this method a machine's behaviour may be made as adaptive as we please, and that the method may be capable of explaining even the adaptiveness of Man.[17]

Alongside F.H. George, Ashby's book had been amongst the earliest of the cybernetic texts that Ascott had encountered. Ashby's aim was that at the heart of early British cybernetics: to further understanding of the complex function of the human brain so that the application of the same systems could be used to create machines with 'adaptive behaviour' – thinking machines. Whereas the analogue machine echoed a situation precisely, the thinking machine would have to be adaptive, deductive and self-managing. This was a possibility that seemed progressively less like fiction and more like the future in the period of heightened technological development following World War II.

The Groundcourse 'system' enclosed the participants and the tutors, drew them together in a manner that blurred the boundaries of production. This mode of pedagogy cannot be limited by a description such as 'class', 'project' or 'exercise' – it was a mass of interconnected ideological concerns approached through experimental techniques as well as creative activities. It was an experiment in the control and mastery of the psychological process of art production; both students and staff were part of the experiment but since the staff had the greater control, it was the students who were the subject. In a series of analytical drawing exercises described by Ascott, the process veered from close observation to free speculation:

> Example 1. Analyse and dissect a section of pomegranate. Discuss with precise drawing its three-dimensional cellular structure. 2. Examine a plant in minute detail; design a new plant based on the principles of growth you have observed. 3. Discuss visually the movements of a hungry, caged lion; then those of a frightened squirrel.[18]

The first exercise in observational drawing asked the student to 'discuss' structure, encouraging experimentation around the relationship of three-dimensional form to the two-dimensional plane. The instruction of using 'precise drawing' made what appears at first to be a simple task more complicated, prompting the respondent to question the veracity of the drawing as an observation and the tension between object and representation. The next two exercises departed from observation and extended into more surreal speculation.

In a surviving photograph of a Groundcourse drawing (Figure 5.1), a student has attempted to describe consumer relationships visually.

From a central core, swirling tentacles of acid colour and pattern open outwards. The exercise was evidently a meditation upon the ways in which information travels from marketing and the kind of processes, mental and otherwise, that might embed the desire for a product in an individual. At the base of the drawing the student had written notes about the relationship between the advertiser and the consumer and what the advertiser must take into consideration in communicating with the consumer. The student writes 'The visual image portrays the relationship between the advertiser and the consumer' and also comments that 'advertisers must take into consideration environmental media'. Many of the Groundcourse students went on to study design; foundation courses served as a shared grounding for both artists and designers, and increasing numbers of students were choosing the various strains of commercial arts that were available in schools of art. A number of the London schools including the Royal College of Art, Hornsey College of Art and the Central School had innovatory courses in graphic design, publicity design and product design.[19] Stephen Willats has recalled that during

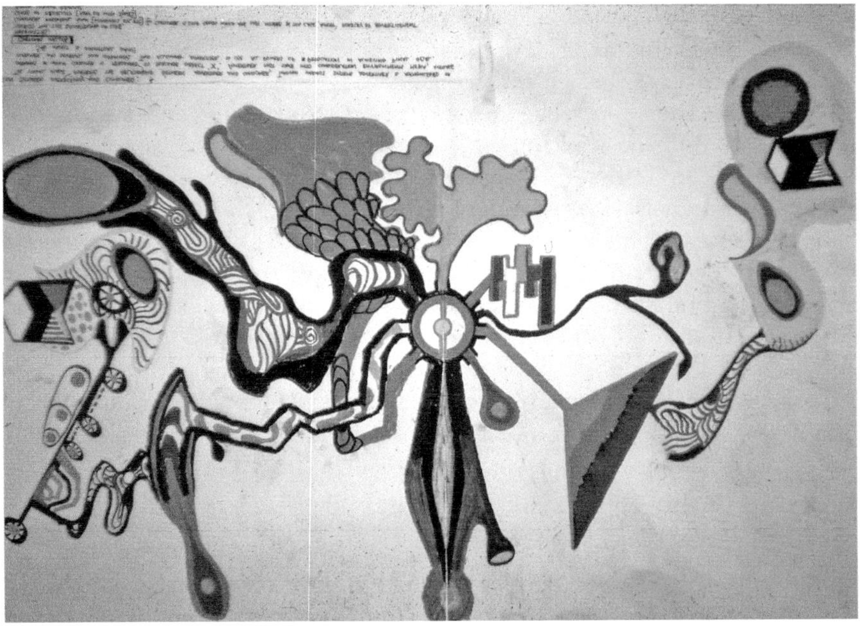

Figure 5.1 Anon (student of Roy Ascott) (1965) *Groundcourse Analytical Drawing*. Ipswich. Photograph © Roy Ascott

his studies, he developed an interest in advertising, because it was '[...] quite advanced in terms of thinking, models of communication and so on'.[20] Ted Powell, who participated in Groundcourse Ealing, went on to a successful career as a commercial design director, and credits his studies with Ascott as a major influence on his subsequent achievements, having instilled an understanding of influence, change and systems of communication.[21]

The drawing focussed on controlled communication and the analysis is directed more towards assessing the way an advertiser must understand their market and target their campaign accordingly. This swirling, psychedelic design dates to the height of the swinging sixties, when the line between counterculture and commodity was often elastic. Ted Powell stayed on at Ealing to study design after Groundcourse. During his diploma course, his animation tutor Peter Green secured work for him and his two friends Nigel May and Bob Miller, creating fillers and tracers on the set of *The Yellow Submarine*. The position was highly lucrative – the students would work until they dropped, spend three pounds on a bed for the night and a Turkish bath before purchasing a fresh set of clothes on Carnaby Street. Every Friday, the director would process through the set with a tall blonde woman dressed as a 'cigarette girl', who would hand out wage packets in place of tobacco.[22] In this sense, this colourful map of advertising strategy is an apt one; psychedelia was certainly commercial.

Psychedelia was also cybernetic; by this I mean that the concept of inducing mind-altered states through using light pulses or biofeedback grew from cybernetic invention, as Pickering has so precisely described.[23] Walter, for example, discovered that flicker strobe lighting induced epileptic fits.[24] In a decade of light shows, swirling lava lamps, dream machines and disco lights, technology paved the way for the countercultural quest for mind-altering experiences. The aesthetics of psychedelic art translated into psychedelic design that adorned not only record covers, but also cushion covers. Flow, the predominate element of this aesthetic, is a visual imprint of a process; most often the pursuit of a higher state of consciousness. The American artist Isaac Abrahams, founder of the first gallery of Psychedelic Art, has mused that it allowed a '[...] turning on to the life process, to the dance of life with all its motion and change'.[25] The close embrace of psychedelic art and psychedelic drugs sometimes excludes these technological contributions to new states of mind. It is also worth mentioning that the quasi-spiritual concept of flow that recurred within psychedelic art was also central to late strains of biological abstraction in Britain.

This exercise also has similarities with exercises incorporated into Basic Design teaching by Hamilton and Paolozzi. Classes at both the Central School and at King's College encouraged students to use the increasingly sophisticated advertising imagery of the late 1950s and early 1960s to collage, reconstruct and question the semiotics of the branded image and the manufactured product. A group of collages originating from King's College and also dating to 1965 (Figures 5.2 and 5.3) used collaged material, drawing and paint to create faces. John Myers' collaged head (Figure 5.2) combined sketchily drawn and cartoonish elements including a muscleman and machine parts with additions including a key, a bottle top and a diagram cross section of an eye. Wilson Bayliss' two collages (Figure 5.3) use glossy advertising imagery such as sunglasses, hotdogs, fruit and balloons to form the features of a face. Wilson's work recalls John McHale's *Machine Made America I* and *II*. The second collage in particular has a dark energy to it, with the gaping mouth of the face painted a solid blood red. The collaged x-ray photograph of a telephone also appears in Richard Hamilton's painting *Desk* (1964).

In both the King's College drawings and the Groundcourse drawing, the language of advertising was both seductive and innately threatening. However, the Groundcourse philosophy of seeking out systems behaviour took the exercise beyond the aesthetics of consumer design, focusing instead on the behavioural triggers used by advertisers. The drawing-diagram implies a relationship between the spiritual principle of psychedelic art and the commercial principle of advertising design, connecting them in systems of influence and control sketched out as a flowing map of interaction. This awareness of psychological manipulation and control was indicative of the broader behaviourism which shaped Groundcourse, and which would also later form the basis of the evolving field of organizational science.[26] There

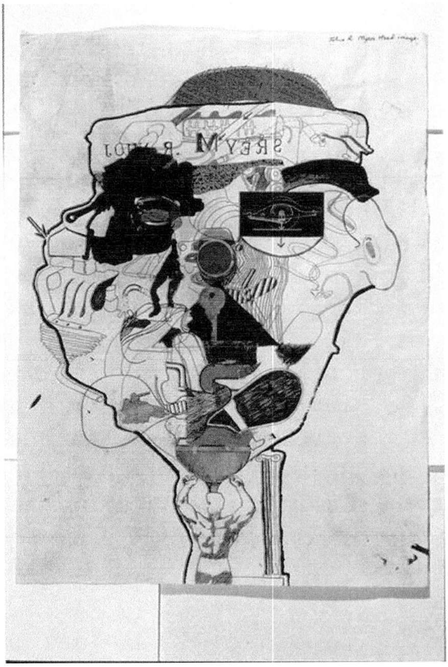

Figure 5.2 John R. Myers (student of Richard Hamilton) (1965) *Head (Image and Anatomy Exercise)* 380 × 510 mm NAEA

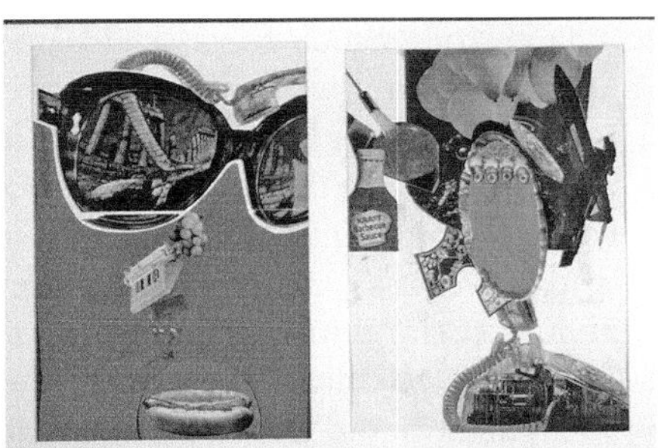

Figure 5.3 Wilson Bayliss (student of Richard Hamilton) (1965) *Head (Image and Anatomy Exercise)* 380 × 510 mm NAEA

were behavioural undertones in this analytical drawing exercise and others – such as those in which students tried to depict the angry or startled movements of animals in different environments as described above.

'Create a world on paper with major and minor structural systems', writes Ascott in his pedagogical notes in 'The Construction of Change', then: 'Show a fault occurring in the minor one; design a repair centre to put it right.'[27] This highly conceptual drawing exercise explored the basic principle of systemic interconnectivity – sub-systems which feed back into the whole. This goes beyond the biological and environmental roots of cybernetics and engages directly with cybernetics as an emerging theoretical model. The students were in this way made to focus on an abstract notion – the interrelationship and mutuality of abstract dynamic systems. This systems thinking might have been an accepted part of the philosophy of science and technology of the age, but placed in the context of the art school studio, the same ideas were novel, alien and distinctly disruptive to any preconceptions of art practice which students had carried with them from the school room. The foundation course had been introduced in order to address the perceived gap between school-level education and the realities of studying art in the age of abstraction. However, as foundation courses were staffed by young graduates like Ascott, they became spaces of invention and innovation rather than the solid grounding in abstract form that had been envisaged by the Coldstream board.[28] Ascott's focus was not bridging the gap through the provision of a set of formal abstract principles; it was building an alternative understanding of the visual arts as a communication system. Disruption was not just an aid for exploring aesthetic decisions; it extended out into other frameworks of communication too.

Mind maps: the brain as a cybernetic system

After the initial process of disrupting and subverting the students' preconceptions had taken place, one of the most curious exercises in twentieth-century visual arts education began. Confusion and uncertainty were intended to be a new beginning, an induced chaos from which form could emerge. The second year of the course was a long, performative experiment that saw the students analyze first their own habits and values before exploring interdependence and systems behaviour. Their remaining time as foundation students took the form of this immersive experiment, in which they were both experimenter and subject, powerful and powerless. Ascott designed the first stage of this project to make students assess their own behavioural traits and to consider ways of overcoming instinctual or habitual patterns of behaviour. This process was, necessarily, analytical, based in the first instance in observing each other through a series of exercises that the students themselves created in order to gain insights into the behaviour of individuals that participated. While this could be approached in any way, game design was one of the most popular approaches, allowing each group of game designers to observe game participants in different circumstances. Brian Eno described creating a game with faceted dice and markers along with his groupmates, before scrutinizing the players and taking notes as they played this game.[29] In this way, the road to self-awareness was prompted by a surreal form of peer review, in which student behaviour in certain playful or performative circumstances was analyzed and criticized. Students were then instructed to create 'mind maps' that described what they had learned about themselves.

These were diagrams which would confirm their likely responses and reactions in different situations. Ascott recalled that:

> Making mind maps to get the students to think about what it was like to be a human being in a given environment [...] I had an ulterior motive which was to really to get them to reconstruct, to re-think themselves. By then we'd worked for a year on getting rid of pre-conceptions then 'what about you?' You are all these possibilities.[30]

One of the most enduring misconceptions about Groundcourse is that this exercise was geared towards the acquisition and enactment of a new, fixed identity when in fact, it was designed to open up multiple possibilities or identities. This was an important principle for Ascott, and it appears at times to have been misunderstood by both former pupils and by historians. Emily Pethick recorded it as an exercise in performing a different self, writing that the exercise resulted in students 'acting out a new personality contra to their own for ten weeks'.[31] The following year, Michael Bracewell quoted Pethick verbatim in his own short account of Groundcourse for *Becoming Roxy Music*, concretizing the idea.[32] In both cases, this notion appears to derive from Ascott's own description of the exercise in 'The Construction of Change':

> Students are set the task of acquiring and acting out for a limited period (ten weeks) a totally new personality, which is to be narrowly limited and largely the converse of what is considered to be their normal 'selves' [...][33]

Ascott did not intend that the new 'personality' would be predictable or pre-patterned; it was habitual behaviour that was to be targeted through limitation. Far from encouraging students to develop fixed new identities, he wished instead to open up their responses and create possibilities. While it was important that the students abandoned their previous reactions, the potential for variety in conceiving a new set of limitations was endless. Ascott reflected: 'It's been misunderstood by some people, that you had to take on an identity, a fixed identity but it wasn't like that at all.'[34] This experiment in identity was intended to make students aware of the possibility of adjusting their own behaviour and thereby altering the outcome of any given situation. The mind maps were essential parts of this pedagogical experiment, a combination of environmental, physical and psychological processes to be explored.

In a student mind map from Ealing in 1963 (Figure 5.4) the student made the following note in pencil: 'Communication structure of 1 member within 7', a note that tells us of the inter-reliance provoked by these exercises. The drawing resembles an electrical circuit, and at its base there are seven soft blocks of colour under a larger block of orange, that look

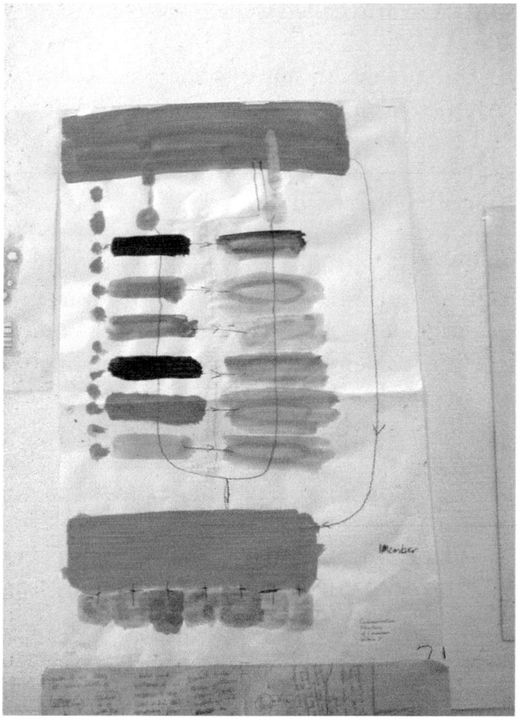

Figure 5.4 Anon (student of Roy Ascott) (1963) *Mind Map*. Watercolour on paper. Ealing

like switches on a fuse board. The amorphous washes of colour together with the tentative diagrammatic lines that connect them, position this image somewhere between the machine and organism. The expressionistic softness of the painting contrasts with the networked communication structure it depicts. Each student retained an individual style and approach despite operating within larger creative entities.

The drawing is a map of interactive participation, of the dynamics of human interaction and individual habits and customs within it, while incorporating symbolic language with a clear cybernetic – or more broadly, technological – underpinning. A second student drawing (Figure 5.5) employs a much more linear network aesthetic, drawn in black, white and limited bright colours on old brown card. Students would make use of whatever materials they could find in the Ealing art rooms and outside, scavenging for scrap. While this is a long established practice for art students, we must also recognize that Britain was still recovering economically from World War II; this generation of students had grown up with rationing and with town and cityscapes marred by war. The instinct to make use of what was available was certainly a prevalent feature of the age, contrasting with the gloss of

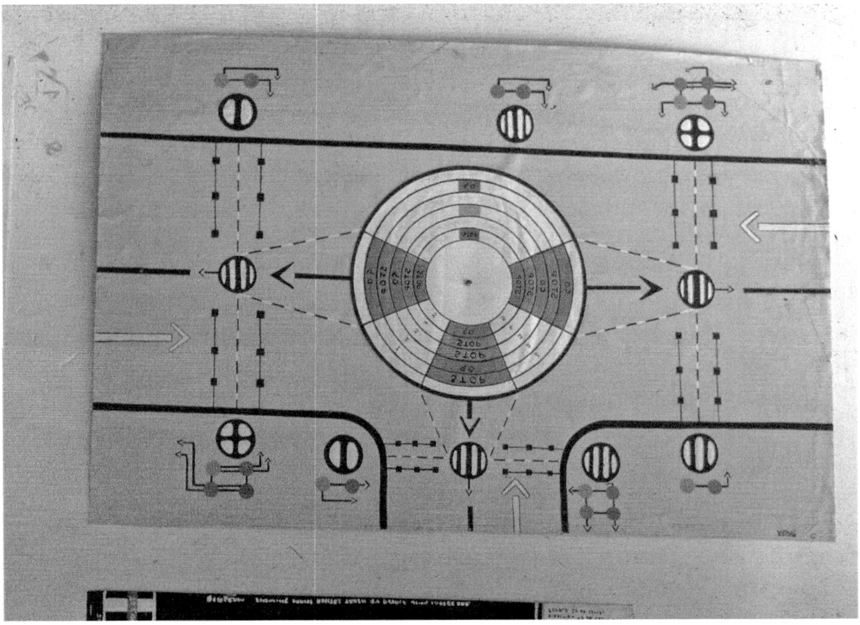

Figure 5.5 Unknown student of Roy Ascott (1963) *Drawing*. Photograph © Roy Ascott

consumerism that countered it and exemplified by the appropriation and collaging of advertising images. While we are familiar with Hamilton and Paolozzi's wistful celebrations and subversions of American and British consumer culture, it is worth noting that the use of reclaimed materials in constructivist work in the UK, within and outside of the schools of art, was partly provoked by a depressed economy.

The drawing has a central wheel which was evidently intended to turn, as there is a pin or tack in the centre that presumably held a spinner or marker. The circle consists of three divided segments each containing the instructions 'STOP' and 'GO' in green and red, in differing orders. Two white segments are numbered 1 to 5 non-sequentially. There are linear arrows radiating from the centre, leading out into a circuitry of cog-like circles. The outer rim of the drawing has seven depictions of coloured dots networked in different configurations by black arrows; these could be possible communication structures between students, or another set of coded possibilities altogether. This drawing then, plays upon chance – the combination of factors selected in the central wheel – and possibility, in the mapping of different coded outcomes.

The diagrammatic quality of these drawings was not a borrowed aesthetic but something that emerged due to the nature of the tasks – in order to express structures, systems and networks, interconnecting lines are a necessity. This networked aesthetic was nonetheless unusual for its time, something that would later cause tensions with an external examiner from the Chelsea School of Art who dismissed the work as 'not drawings [...] just a lot of bloody diagrams'.[35] These drawings were not figurative but nor were they abstract; they were depictions of relationships and as such, they were to prove difficult to assess. As noted, in the 1960s the gap between the old and new guard in British schools of art was considerable and by extension, this level of misunderstanding around production values in new art and design courses was all too common.

The mind maps were as biological as they were technological, representing living systems of interaction. The notion that cultural processes could be perceived as, or compared to, a kind of complex ecology evolved alongside early cybernetics, principally in anthropology. In 1955 the anthropologist J.H. Steward wrote: 'Analogies between cultural and biological evolution are also alleged to be represented by two attributes of each: first, a tendency towards increasing complexity of forms and, second, the development of superior forms, that is, improvement or progress.'[36] His interwar work on the notion of cultural ecology was eccentric for its time but it gained greater contemporary relevance in the post-war years. In *Theory of Cultural Change: The Methodology of Multilinear Cultural Evolution*, he presented an extended argument which compared social and cultural structures to biological form:

> Just as simple unicellular forms of life are succeeded by multicellular and internally specialised forms which have distinctive kinds of total organisation, so social forms consisting of single families and lineages are succeeded by multifamilial communities, bands, or tribes, and these, in turn, by state patterns, each involving not only greater internal heterogeneity and specialization but wholly new kinds of overall integration.[37]

The concept of cultural/creative ecology has in the intervening years become familiar to the point of overuse, serving as a convenient metaphor to express the complex reactions and interactions that lead to cultural development. Noting that cultural evolution had long been abandoned as an old-fashioned concept, Steward positioned his work against other anthropological studies which utilized the same methodological approach: principally those of V. Gordon Childe and Leslie White.[38] He described how the previous two decades had seen a resurgence of interest in cultural evolution which had previously been a much-maligned idea.[39] However, his book can be better understood in the context of post-war overlaps between cybernetics, biological systems and the humanities – his view of pattern-generation reached from the organismic to the social. When Ludwig von Bertalanffy argued for a General System Theory, he outlined the fact that systems behaviour was evident in all realms of science, nature and culture.[40] He argued that for many years, science had focused on the '*world as chaos*', but that it was time to seek out the '*world as organization*'.[41] He wrote:

> [...] there exist models, principles, and laws that apply to generalised systems or their subclasses, irrespective of their particular kind, the nature of their component elements, and the relationships or 'forces' between them. It seems legitimate to ask for a theory, not of systems of a more or less special kind, but of universal principles applying to systems in general.[42]

In a sense then, Steward's off-beat ecological anthropology was simply a systems approach. It was certainly evident that the discipline of anthropology was changing in response to post-war culture; the anthropologist Margaret Mead, for example, applied cybernetics to her vision of the future of the discipline. She and her husband Gregory Bateson were amongst the original participants in the Macy conferences, a series exploring cybernetic thinking in the USA.[43] Bateson later published his collected essays under the title *Steps to an Ecology of Mind*.[44] In his introduction he stated: '[...] such matters as the bilateral symmetry of an animal, the pattern-arrangement of leaves on a plant, the escalation of an armament race, the processes of courtship, the nature of play, the grammar of a sentence, the mystery of biological evolution, and the contemporary crisis of man's relationship to his environment, can only be understood in terms of an ecology of ideas as I propose.'[45] Bateson argued that there was a need to connect behaviour with natural pattern, given the developments made in the areas of cybernetics and systems theory. He intended to build '[...] a bridge between the facts of life and behavior and what we know today of the nature of pattern and order'.[46] Bateson argued that systems theory extended into the realm of behaviour; behaviour too, was produced by an intricate system of interacting elements.

The question of how anthropologists might contribute to a technologized future led to a reassessment of the problematic polarising and historicizing of cultures both living and dead as a result of existing anthropological methodology. The anthropological past was then reassessed, networked, reconnected to the living world. A surprising trend for using science fiction as anthropological teaching material emerged in the US and the UK.[47] In this way, anthropology addressed the fluctuating convergences of technology, cybernetics and processes of cultural development that characterized the era. Network thinking extended beyond the boundaries of cybernetics, serving as a useful model to visualize a world in which communication technologies and increasing interconnectivity were changing relationships between people. Ascott's students were presented with a model of practice in which what they produced and achieved was treated in relation to what others produced and achieved, in opposition to the long-standing celebration of individual achievement in visual arts training. In his 1967 essay 'Behaviourables and Futuribles', Ascott had reflected that 'Now we see that the world is all process, constant change, we are less surprised to discover our art is all about process too'.[48] Behaviourist art had 'two principle aspects – the biological and the social'.[49] Like Bateson then, Ascott recognized that patterns of human behaviour were essentially cybernetic; systematized and interdependent. He also recognized that both individual and collective behaviour were systems and could be analyzed as such.

Calibration

While the Groundcourse mind maps were essentially exercises in analysis and behaviour, another exercise took the process of self-awareness into a more interactive realm. The students built what Ascott called 'calibrators': devices that would allow the user to read their instinctive responses against other possibilities, actively 'calibrating' their behaviour in accordance with a given situation. The calibrators '[...] ranged from handheld little devices to quite complex structures', designed and built by each student.[50] Ascott used the term calibration to express the concepts of self-measurement and self-adjustment. Each calibrator had to contain a range of options so that they could be adjusted analogously with the situation, including environment, action and reaction. They were specific to the environment of the art studio, and as such, the immediate environment included choices of art materials and tools 'which one might use creatively'.[51]

Ascott recalls that forms of behaviour in a studio context such as '[...] walking, thinking, handling, sawing, joining, seeing, singing, talking [...]' were often listed at this stage. For each individual, all options would be 'severely limited', in order to create the organismal interdependency that Ascott intended the students to maintain in their groups.[52] From the examples of calibrators and from the memories of students who took part in this exercise though, it is evident that other more personal modes of calibration also emerged. Some students explicitly calibrated ranges of emotional or personal responses to situations, from anger to vulnerability. A dual understanding of calibrating a situation developed – the purposive calibration of environment and action that Ascott described, and the emotional and reactive responses that some students considered. This is a good indicator of how controlled models of pedagogy such as Groundcourse strayed onto psychological territory, perhaps because self-reflection and insecurity are so endemic in the early years of art training. Given that many pedagogical approaches in post-war art education in Britain used shock and uncertainty to provoke responses from students, it is no surprise that self-awareness, performativity and emotional reflection became increasingly common elements in the resulting work.

Calibration, then, offered the students a chance to reflect upon their own decision-making processes and the habits, emotions or physical limitations that shaped them. The term 'calibration' bestowed the process of self-scrutiny with a mechanical, objective distance, systematizing responses and creating a spectrum of possibility. The idea of calibration was intrinsically mechanical, a term that was used at the time principally in engineering contexts in reference to a machine's functionality. To calibrate a machine was to make sure its settings reflected the information it needed to function; for example, the increasingly systematized operation of World War II weaponry needed to be updated with information such as wind speed and direction, tides, distances and altitude. Calibration is preparation; it involves taking in information and adjusting course accordingly. This exercise was intended to make students reassess process and outcome and to move towards a more self-aware mode of practice.

One calibrator was evidently designed to make the student use an invented language (Figure 5.6). This calibrator displays symbols, letters and numbers in concentric circles, with the outer three rings containing an invented alphabet. The English alphabet and the numbers 1–26 in the centre could evidently be aligned with the invented symbols by moving the tab, in order to interpret words from, or into, this mysterious alphabet. Pete Townshend remembered that after the analysis of his behavioural traits, he and his group had limited his use of language to hamper his tendency to talk incessantly.[53] The invented language made his communication slower, more considered and also less likely to be received easily by an audience. He had become reliant on the attention of others and this new manner of communication limited his speech to short and considered contributions, delivered in code.

This mind map certainly operated on the principle of interpretation, its revolving circles operating on the principle of the slide rule. It bears a striking resemblance to early cipher devices such as Plett's Enciphering Machine. Just like Plett's code-making device, the mind map had a spinning cog at its centre. The creation and deciphering of codes had great contemporary relevance; coded communication had been one of the most important elements of warfare during World War II and cracking codes drove the secret activities of the military during the Cold War. Code-breaking is also the most perfect example of the technological transition that had taken place during

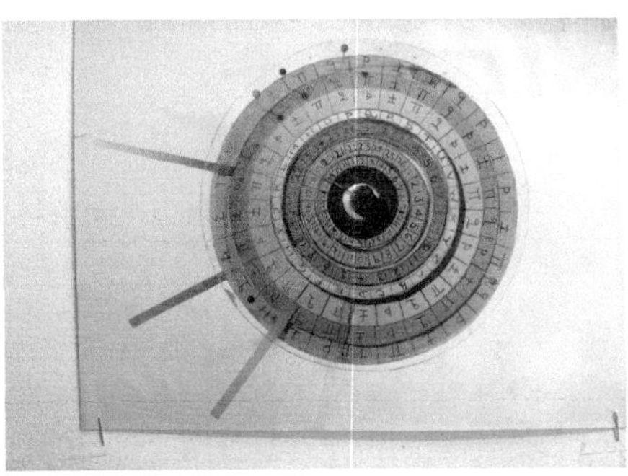

Figure 5.6 Anon (student of Roy Ascott) (1963) *Behavioural Project Calibrator*. Ealing. Photograph © Roy Ascott

the war; machines could now perform calculations that were too complex for the human mind. In this sense, then, slide devices such as code breakers symbolize an extension of the human ability to calculate and assess. Another example of commonly used slide rules were those used by pilots; handheld computers with speed and height on one side and a tactical grid on the other with a vector triangle of heading, wind and track (Figure 5.7). Using the slide rule, a pilot can judge and measure position and estimate course towards a target. These small calculators are still used in the training of private pilots today; they offer an analogue backup in the event of technological failure and a way to make sure that the trainee explicitly understands the calculations that have for so long now been digital.

The pilot operated the slide rule by using a previous position to calculate a current position from the known factors of speed and direction; in this way the pilot manipulated the object to be analogous to the situation. They are palm-sized, intimate and accurate, allowing a complex calculation to be performed quickly and simply and they were used by thousands of pilots during World War II. The pilot would move towards a target based on previous intelligence, in that each calculation was performed either before or during flight. The slide rule was entirely dependent on interaction and incapable of taking readings. It produced information from the input of information. Despite the technological development that had led to analogue systems that self-adjusted, the slide rule had a reassuring analogue reliability. Radar information could be transmitted to an analogue computer where information was continuously updated to match the continuous flow of information, and projected onto a screen. In this transient stage between the analogue slide-rule and the technical possibilities of radar, the slide rule itself was still safely in the pocket of every pilot. The first slide rules were invented in the seventeenth century, in response to the work on logarithms done by John Napier. Simple manipulations of the slide rule resolve

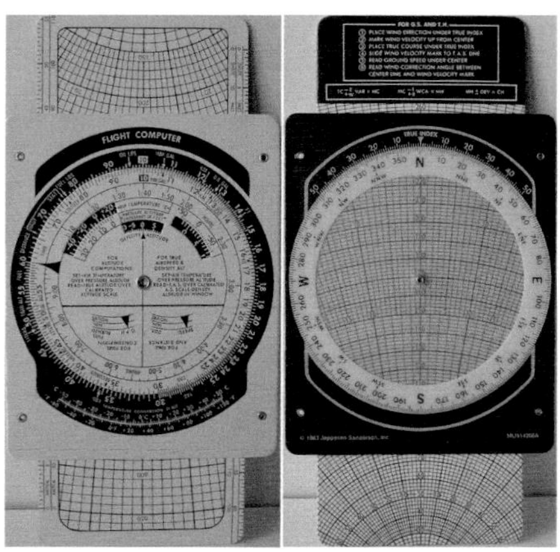

Figure 5.7 Pilot's Slide Rule E6B. Photograph © Dave Paige

calculations, such as the vital measurements that a pilot needed to chart his course. Unlike the uncharted virtual spaces opened by new technologies, the connection between the analogue object, the pilot and the wider environment was precise, pure and complete. In essence, then, the slide rule was an emblem of systems interaction; it depicted the many factors that contributed to the movement of an aeroplane across the sky. Its seductiveness lay in its ability to precisely reflect change.

In Ascott's pedagogy notes, he describes these calibrators as a tool to 'read off responses' to a given situation.[54] To operate a slide rule, the user must first know basic information such as, in the case of the pilot, point of departure and speed, which can then be measured against readings taken from the new situation. In the case of a personality calibrator, the student would be able to measure their new response to a given situation against the basic information of their natural responses, so that they essentially performed an alternative response. The calibrators were part of a web of image, language and communication which the student artists had to negotiate. Many of the Groundcourse calibrators employed the style of slide rules, in that they were built around moving elements that could be aligned in response to information. One such calibrator created at Ealing in 1963, marked 'Calibrator of Human Characteristics', is a hand-held circular slide-rule, built in card (Plate VI). It operates around moving tabs and dials and it is structured around detailed information. It lists a range of human traits, characteristics and habits which can be adjusted with the sliding mechanism.

The bottom tabs list the following possible traits:

> ambitious, daydreams, suspicious, nagging, prank, aggressive, untidy, mental, eats a lot, romantic, drinks, deaf, willing, pompous, superficial, versatile, vulnerable, eccentric, slow, smooth, placid, vague, vigorous, hypersensitive.

Operating a personality calibrator requires self-analysis, with the hopeful outcome of greater self-knowledge. It is a vulnerable act, confronting the weaknesses or patterns in behaviour and analyzing what would happen if another path were chosen. This calibrator operated analogously to a student's reactions – an aggressive response would have to be noted by sliding the wheel and aligning with an alternative. It was a fluid measure – taking in immediate responses and then readjusting to the new course, which the student would set and attempt to follow. The calibrators were transformed through use in that their own state changes in line with their operator; each calibrator was made and remade by the intentions, beliefs and values of the individual who used it. Ascott's own interest in the analogue concept played a part in this exercise, in which evolutions of object, environment and viewer/participant took place.

The remaining examples of calibrators from Ealing and Ipswich highlight that this exercise hovered between the formal and the personal, with unanticipated and emotional responses to an analytical exercise. At the same time in visual art training, the last vestiges of abstraction were giving way to a diaspora of conceptual approaches. Groundcourse was certainly unusual in that much of the curriculum was directed towards self-analysis, both group and individual, as opposed to the longstanding touchstones of material and process. It is certainly worth noting that models of visual arts education that focus on individual, and often self-generated, development have dominated in British schools of art since the decline of abstraction, making the early 1960s an important transitional moment. In this sense, the self-consciousness that emerged amongst the student body at Ealing and Ipswich was a product of a subtle shift of power, placing the power of decision-making into the hands of the students themselves. It should also be noted that self-consciousness in itself produced more performative modes of working from the students, blurring the line between learning process and product. In a surviving photograph of a calibrator in use (Plate VII) the figure pauses, a gun-shaped calibrator in one hand and a pen in the other. Sitting beside a screen of painted wooden flags, he looks absorbed, as if decisions have a heavy weight in this behavioural experiment. In the bottom right of the image, a paintbrush still stands in a bottle of yellow paint that matches the flag in the midground. It was an evolving, sculptural environment.

The artist Stephen Willats' sculpture from the Groundcourse years is an interesting example of how the same issue of process and product extended into what might be described as object performances during Groundcourse. Willats studied with Ascott at Ealing and he was later invited by Ascott to teach some classes at Groundcourse Ipswich. Like Ascott, Willats sustained an interest in cybernetics and systems theories throughout his career. During his time at Ealing, he was present for Gordon Pask's talk, and he later had an office job at Pask's firm Systems Research.[55] In the catalogue for his 1979 Whitechapel exhibition, *Concerning Our Present Way of Living*, Willats described the work as follows:

> This work was one of a series of six constructions which together formed a learning system. Each work in the system presented the participant with a set of variables held within a fixed framework that could be manually changed into a self-determined order. The set of variables increased from work to work, moving from a wall construction where only small changes of hinged planes were possible to the omni-directional possibilities of the large red cube, surrounded by sets of smaller cubes which could be plugged in on any of its six faces [...].[56]

The sculpture was the third of a series of six, moving from wall pieces through to cubes such as this one, each of which were designed to be manipulated by the viewer. Moreover, each viewer had to record the changes he or she made to the works of art: 'Changes were to be noted by the participant on an accompanying form, displayed on the wall alongside the work. The series was designed to involve the participant in making relationships between perception, decision-making and self-determined behaviour.'[57]

This sculpture relied on cumulative qualities of information formed of the manipulations made by previous viewers. In addition, the 'system' of the six sculptures became an analogue for the decisions and actions of these viewers. The systemic set-up of the sculptures, charts for note taking and sequentially developing forms together created an extended interactive environment. This series of sculptures is very much in the vein of Ascott's contemporaneous analogues, although with a more formal aesthetic. Willats' *Colour Variable No. 3* has the bright simplicity of a puzzle, a toy or a game, in keeping with the elements of game play that formed a recurrent part of the Groundcourse curriculum. The note-taking element of the 'performance' of these sculptures is a prelude to Willats' later practice, in which information gathering in the style of the social sciences was one of his most established techniques. The 'information' collected from *Colour Variables* was a transcript of decisions, a map of the many visual routes that could be taken as the sculptures led the participant from limited choices to more complex ones. Willats described how the:

> [...] structure was made from repetitive elements it could arrive at any particular form, which would be simply the point at which a person stopped adding fresh elements. As all elements were considered to have the same possibilities within the structure, any configuration that they ended up in was to be considered valid.[58]

In 1962–3, Willats had completed a series of drawings titled *Organic Exercise Series Two* which he remembers '[...] as studies for a group of participation sculptures that I made in the early sixties, which expressed my concept of the "democratic structure"'.[59] These drawings responded to that contemporary architectural form that was the tower block, exploring the symbiotic external and internal structures. The sculptures that Willats made in this period were built from a variety of materials, including wood, brick, cement and stone blocks.[60] With an echo of the interactive sculpture produced at King's College, they operated in different ways, including 'hinged planes that could be moved about' and '[...] others which were to be built up bit by bit from basic variables, the next participant starting from the form left by the previous participant'.[61]

Considering the differences in interactivity between King's College and Groundcourse then, it is evident that the change that had taken place was, at this point in time, more conceptual than material. While students worked with similar base materials, the Groundcourse students were working with a new concept of the work of art as unstable, continuously changing through interactivity. The analogue concept that Ascott applied within the Groundcourse curriculum resulted in constant interactivity, coupled with information-gathering practices. This created a circular process where research, making and displaying/performing led back to more research. Given the technological origin of the analogue machine, this flow of 'information' as a conceptual idea doubles back to its military origins. As if in the military bunker, students took readings, dials, displays, cross-referencing data and updated the flow of visual information. This curious collision of Cold War technologies and visual arts training offers us an insight into what was a much larger and more complex phenomenon: that is, the impact that communication technologies originating in World War II had upon the gestation of conceptual art practices. The work that Ascott and his contemporaries were producing and the related pedagogical models offer perhaps the most direct evidence of an influence that dispersed and was absorbed widely.

While Willats' work had a high level of finish, many of the interactive sculptures created during the course were elementary, even crude in their construction. Built from elaborate loops of paper, naïve structures in wood and card, offcuts and junk, the sculptures were built with interaction and measurement in mind. Often such analogue sculptures were the prescribed outcome of a particular project or exercise. In Figure 5.9, students manipulate analogues, taking readings from tabs and dials. When the analogue became a student exercise, an interesting phenomenon occurred. The formal playfulness of the concept developed a technological aesthetic that recalled cipher-breakers and other early computers. Aside from the legacy of the very specific employments of analogue technology during the war, the Groundcourse analogues exhibited an aesthetic and conceptual connection to the sophisticated technologies even though they were built not from metal and glass, but from paper, card and wood. John Bonehill remembers: '[...] everything we made was made from paper, string, rubber bands, clockwork, electric motors with huge batteries, hot air, explosives, paint, pencils [...] you get the idea, no assistance from computers because they hadn't been invented yet on a scale smaller than a post office.'[62]

The likeness to cipher breaking was unspoken and unrecorded but it remains visually clear in the pair of photographs. It was 1963. The body of students was predominantly seventeen or eighteen years old – the immediate post-war babies. Their childhoods had taken place in the years of recovery, their food still rationed until 1954. The first phase of the Cold War spanned their early lives, characterized by secrecy, technological advancement and watchful tension. The previous year, the students had continued to attend college as the Cuban Missile Crisis reached its zenith. Pete Townshend recalls that:

> On the critical day in October 1962 I walked into college absolutely certain that my life was over; why was I even bothering to attend class? When the end didn't come I was glad not to have been one of those who had panicked, wept or chattered compulsively until the good news was announced.[63]

The psychological and emotional impact of the tense military negotiations taking place on a global scale manifested itself silently – something written clearly into the visual language of the analogues.

In these machines of card and string, elastic bands and offcuts, a space of performance opened up. These were machines and they were not; they operated somewhere between purpose and play. Perhaps it was the camera directed at them, or perhaps it was the focussed manipulation and note-taking. The art school studio appears staged, driven by self-awareness. Like Willats had arranged for his *Colour Variable* series, the students were making notes from the analogues, manipulating them and recording their 'findings'. The object, its immediate environment and the people around it were drawn into the work. The course was in this light a network of students, staff and objects within defined environments. The objects – games, analogues, calibrators and costumes – operated like switches, measures or obstructions to the flow of activity.

While it is tempting to read Ascott's analogues as objects with agency, living, changing and influencing, they were closer to props. They reflected an abstract flow of ideas that was in itself alive – because it was human. Ascott understood from the outset that communication technologies were simply the means to extending and transforming human communication – a computer is merely a shell but the flow of information it allows is alive, transformative and powerful. Early media art is often subjected to a troublesome prejudice against – or misunderstanding of – its aesthetic qualities. In Frederic Jameson's seminal account of postmodernism, he lingers for some time on the fact that contemporary machines had little 'emblematic or visual power', in comparison to 'the turbine', 'Sheeler's grain elevators or smokestacks', the 'baroque elaboration of pipes and conveyor belts' and 'even the streamlined profile of the railroad train'.[64] He found modern technology to be bland and featureless, describing 'the casings of the various media themselves, as with that home appliance called television which articulates nothing but rather implodes, carrying its flattened image surface within itself'.[65]

It is certainly true that many of the machines and components that were employed by artists working with technology in the 1960s and 1970s were visually unappealing, designed for specific technical function and not for visual pleasure. However, when experiencing a piece of media art, just as in watching a television set, the machine is a tool for an experience. It is apparent that one of the primary reasons that Ascott's work has not had the scholarly attention it deserves to date is simply that people are not interested in, or distracted by, the 'shells' – computers, mechanical detritus, the memories of performances with trailing wires and messy, unpredictable machines do not exhibit well and do not reproduce well in books. While examples of the technologies that Ascott employed still exist, they are inert. To get past this problem, we merely have to analyze the content not the computer, the flow rather than the form.

To illustrate this, we can turn to the student calibration devices. Given that they disrupted or redirected the decision-making process of their users,

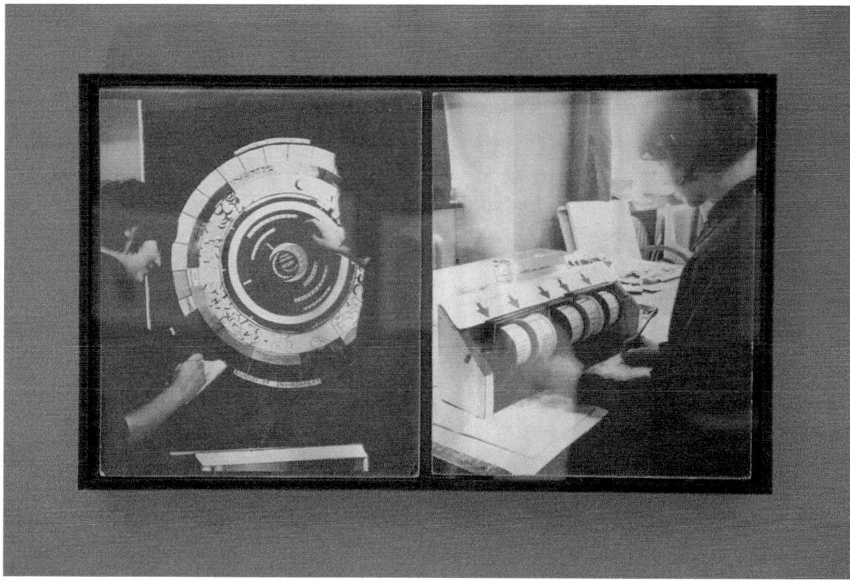

Figure 5.8 Roy Ascott (1963) *Behavioural Project Ealing*. Photograph © Roy Ascott

they quite naturally prompted the students into a mode of performance. The material form of the calibrator mattered only inasmuch as how fit for use it was – like game design, calibrator design was a way of prompting a set of decisions or encounters. This was a self-conscious and open-ended project that required students to make decisions in a concentrated way. It was also immersive, in that there was no dividing line between the individual and the making process. In this sense, it veered into the territory of play or performance, and it was often a play upon code and meaning. The performative manipulation of these analogues has a definite echo of code-breakers at work, particularly in the right-hand image in which the analogue sculpture so clearly resembles a real cipher machine (Figures 5.8 and 5.9).

In the 1960s, the extent of the contribution made by code-breakers during and since World War II would have been unknown to staff and students as it was still shrouded in secrecy. Despite this, the role of the cryptographer in decoding information and increasing national security was a familiar one, given that ciphering equipment had been used long before the war. After the war though, the ghost of the cipher machine manifests itself as several contemporary concerns – *information*, *code*, *communication*. Ascott had absorbed several theories and ideas relating to the flow of information as a disembodied, conceptualized entity.[66] The ideas that occupied him as an artist extended into the laboratory of the art school studio. As Ascott's ideology dispersed and transformed through the work produced by students and staff, it is interesting that such explicit visual links to communication technologies emerged from an abstract set of ideas that had been distilled from technology.

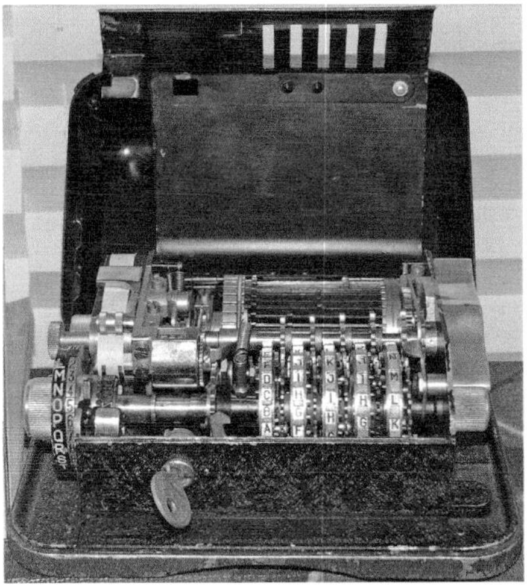

Figure 5.9 Hagelin Cipher Machine. American. World War II

Taking in the fact that communication technologies still held primarily military associations, this strain of interactive sculpture in Groundcourse presents an interesting issue. While the field of cybernetics had its roots in World War II, it was very much future-facing and disassociated from the weaponry that had prompted several scientists and theorists to consider systems behaviour in man and machine. Likewise, the field of behaviourism segued into business and social applications that disassociated it from wartime logistics and control models. Despite this, it was evident that for students, systems thinking – even in the abstract space of the art school studio – retained military associations. Given that behaviourism and more broadly, systems of control, were foundational elements of the course, it is evident that for the students at least, control was not an abstract concept. It carried with it all the associations of its origins. Calibration then, was an exercise in self-analysis – but this was a means to understanding the place of the individual in a network of communication. Groundcourse opened out gradually from the individual to the collective, establishing, in stages, a systems view of art production that took in creation and reception. The collective method of working also meant that any works of art produced were chiefly collaborative, disrupting the focus on individual assessment upon which art training was built.

Organisms, interconnection and group dynamics

Calibration formed a preparatory level of engagement and self-analysis before the course proceeded with its most extensive project; the students were placed into groups. These groups were, in Ascott's eyes, 'organisms'; they operated as a system of individuals that collectively achieved their function:

> They form groups of six. These sexagonal organisms, whose members are of necessity interdependent and highly conscious of one another's capabilities and limitations, are set the goal of producing an *ordered entity* out of substances and space in their environment.[67]

Each organism was reliant on its relationship with its environment. Every individual had a role to play in their organism, and the calibration process of analyzing behaviour held an important place in deciding what, precisely, that role would be. The term organism encompassed the fact that the groups were living entities, working to a mutual goal. Beyond this, it reflects the fact that systems, living or technological, have *intention*, surely one of the most interesting recurring traits charted by the first wave of cyberneticists.

The function of the organism took precedence over the individual. Indeed, each individual had their range curtailed in some way, particularly with regards to their predominant character and behavioural traits. It became apparent under calibration and observation if a student was shy, aggressive, impatient, talkative, lacked concentration or preferred speaking to making. Each student had to avoid the habitual behaviours which this process of group analysis had detected – not for an hour or a day, but for the rest of the term. Ascott described this interconnectivity in relation to game play and calibration:

> [...] they would have three kinds of elements to deal with; first of all we would set them a project what we want you to do is to invent and design and build and play a game and to do it you'll be in groups of six – five is a better number but it was six – the thing of it is you could only do what this thing called a calibrator allowed you do on this calibrator you would divide up room size, number of people, whatever against running, standing sitting, no eyes, no legs, no ears, no speech and that would be like behaviours, environments, materials, sand, rope, paper [...][68]

Within these groups, students would be set tasks which related to their personality mind-mapping; more specifically, they would be given responsibility for something which they would find difficult:

> The student who thinks himself 'useless' with, say, colour, machine tools, or objective drawing may find himself with sole responsibility for these things in his group. The shy girl must act out an easy sociability; the aggressive youth must become cooperative. One student may be limited to transporting himself about the school on a trolley; another may not use paper, numbers, or adhesive substances.[69]

There were two levels of intervention here – firstly pedagogical, in that the students had to work at what they perceived to be their weak points, and secondly behavioural, in that they had to try to overcome dominant personality traits. This was a pressurized task to the extent that they were contributing to a larger goal, so the success of each group relied upon the success of its composite members.

The student Ascott mentioned transporting himself around the school on a trolley was in fact Brian Eno, although there is some confusion about this matter. In Shanken's account, he named Pete Townshend as the student described in Ascott's exercise, but in Michael Bracewell's book *Re/Make Re/Model: Becoming Roxy Music*, Eno reminisced about the exercise:

> [...] I wasn't allowed to move. I was very energetic, and so I had to sit on a goods trolley until somebody moved me. Also, because I tended to like making plans but was hopeless about doing anything, I became the person who had to make things – actually build things. When something needed to be made, they would wheel my trolley into the workshop and I had to make it – and this went on for the whole first term.[70]

Given that Eno is quoted directly on this matter, we can safely attribute this particular experience to him at Ipswich. As previously noted, Townshend was limited to speaking through a created alphabet to curtail his constant talk, although he does also recall being forced to move around in a cart built of orange crates.[71] There are certain misunderstandings in Eno's recollection of his experience in comparison to Ascott's intentions in that the pedagogy was not designed to limit the students, but rather to stimulate them to find new solutions and modes of behaviour, an intervention not into creative process, but into self-awareness. Ascott's hope was that the students would realize that *self* was an unlimited state; in fact, self was process. Self as process was subject to intervention, disruption and revolution from the wider environment that consisted of interconnected individuals. While there was at times distance between Ascott's intentions and how students interpreted the meaning of activities they were set, the principle that we have choice in our behaviour still came across clearly. Eno's words offer an insight into the dissonant space between pedagogy and pupil in which the outcomes of a model of teaching can produce surprises, deviations and unexpected successes. Groundcourse produced self-conscious students – not makers but participants who learned through a distinctive pedagogy of experimental play.

Notes

1 Pete Townshend (2012) *Who I Am*. Glasgow: Harper Collins, 256.
2 *Ibid*., 256.
3 *Ibid*., 50.
4 Several of the Basic Design pedagogues taught summer schools in the UK for art teachers and artists, in the same model that they used for first year DipAD students. This included Pasmore, Tom Hudson and Harry Thubron.

5 Kate Sloan *Interview with Roy Ascott* (September 2012).
6 Exposed to the fierce competition amongst artists in 1960s London, Ascott wanted to set out his position: he decided he was to be 'a man of ideas'. Kate Sloan *Interview with Roy Ascott* (November 2015).
7 John Bonehill (2011) Email exchange with the author.
8 *Ibid.* (2011).
9 John Bonehill (2011) Email exchange with the author.
10 Roy Ascott (1964) 'The Construction of Change'. *Cambridge Opinion 41 (Modern Art in Britain)*, 37–42. Reproduced in Ascott & Shanken (2002) *Telematic Embrace*, 97–107.
11 Ascott (1964) 40.
12 Ascott (2002) [1964] 105.
13 There are several drawings and photographs relating to this strain of Basic Design teaching in the collection of the National Arts Education Archive, based at the Yorkshire Sculpture Park.
14 For an overview of the Independent Group, see Anne Massey (1995) *The Independent Group: Modernism and Mass Culture in Britain, 1945–59.* Manchester: Manchester University Press.
15 Ashby (1960) 40.
16 Ascott (1964) 41.
17 Ashby (1960) 1.
18 Ascott (1964) 41.
19 For an account of this in relation to the RCA, see Alex Seago (1995) *Burning the Box of Beautiful Things: The Development of a Postmodern Sensibility.* Oxford: Oxford University Press, 26.
20 Stephen Willats quoted in: 'Stephen Willats in Conversation with Christabel Stewart' (2014) Event at Raven Row during the exhibition *Control. Stephen Willats. Work 1962–69.* Transcript accessed online on 1202/2017: http://www.ravenrow.org/texts/55/
21 Kate Sloan *Interview with Ted Powell* (by telephone, 2017).
22 *Ibid.*
23 See Andrew Pickering (2010) *The Cybernetic Brain: Sketches of Another Future.* London & Chicago: University of Chicago Press.
24 *Ibid.*, 77.
25 Isaac Abrahams quoted in: Stanley Krippner (2017) 'Ecstatic Landscapes: The Manifestation of Psychedelic Art'. *Journal of Humanistic Psychology*, Vol. 57. No. 4, 415–35, 423.
26 One of the foremost authors in the field was the cyberneticist Stafford Beer, who presented an explicitly cybernetic model for organizational management. See Stafford Beer (1959) *Cybernetics and Management.* London: The English Universities Press Ltd, and (1972) *The Brain of the Firm.* London: Penguin.
27 Ascott (1964) 41.
28 William Coldstream (chair) NACAE 'First Report of the National Advisory Council on Art Education' (*The First Coldstream Report*), 1960.
29 Kate Sloan *Interview with Brian Eno* (May 2016).
30 Kate Sloan *Interview with Roy Ascott* (2012).
31 Emily Pethick (2006) 'Degree Zero'. *Frieze Magazine.* Issue 101, September. Available online at http://www.frieze.com/issue/article/degree_zero/, accessed 1 October 2011.
32 Bracewell (2007) 196.
33 Ascott (1964) 105.
34 Sloan (2012).
35 Kate Sloan *Interview with Roy Ascott* (2012).
36 J. H. Steward (1955) *Theory of Cultural Change: The Methodology of Multilinear Cultural Evolution.* Illinois: University of Illinois Press, 12.

37 *Ibid*. 16.
38 Steward (1955).
39 *Ibid*. (1955) 12.
40 Ludwig von Bertalanffy published the first fully articulated model for a General Systems Theory, based on earlier articles. See: Ludwig von Bertalanffy (1969) *General System Theory: Foundation, Development, Applications*. New York: George Braziller.
41 *Ibid*., 187–8.
42 *Ibid*.
43 For a discussion of the Macy conferences and the origin of cybernetics, see N. Katherine Hayles. (1994) 'Boundary Disputes: Homeostasis, Reflexivity, and the Foundations of Cybernetics'. *Configurations*, Vol. 2. No. 3, 441–67.
44 Gregory Bateson (1972) *Steps to an Ecology of Mind*. New York: Ballantine.
45 *Ibid*., xv.
46 *Ibid*., xxiv–xxvi.
47 See Samuel Gerald Collins (2003) 'Sail on! Sail on!: Anthropology, Science Fiction, and the Enticing Future' *Science Fiction Studies* Vol. 30. No. 2, 180–98.
48 Roy Ascott 'Behaviourables and Futuribles'. Reproduced in Ascott & Shanken (2002) 157–61, 157.
49 *Ibid*., 159.
50 Ascott & Sloan *In Conversation*. The Henry Moore Institute, Leeds, 8 March 2017.
51 *Ibid*.
52 *Ibid*.
53 Pete Townshend (2012) *Who I Am*. London: Harper Collins, 50.
54 Roy Ascott (1964) 'The Construction of Change'. *Cambridge Opinion 41 (Modern Art in Britain)*, 37–42. Reproduced in Ascott & Shanken (2002) *Telematic Embrace* 97–107, 105–6.
55 Willats (2014).
56 Stephen Willats (1979) *Concerning Our Present Way of Living*. Whitechapel Art Gallery & Stedelijk Van Abbemuseum, Eindhoven, 14.
57 *Ibid*. (1979) 14.
58 Stephen Willats (1988) *The Tate Gallery 1984–86: Illustrated Catalogue of Acquisitions Including Supplement to Catalogue of Acquisitions 1982–84*, London: Tate Gallery, 297–8.
59 *Ibid*., 298.
60 *Ibid*., 298.
61 *Ibid*., 298.
62 John Bonehill (2011) Email exchange with the author.
63 Pete Townshend (2012) *Who I Am*. London: Harper Collins, 50.
64 Frederic Jameson (1991) *Postmodernism or, The Cultural Logic of Late Capitalism*. London: Verso, 36–7.
65 Jameson (1991) 37.
66 See 'Analogue' chapter for a full discussion of Ascott's engagement with information theories.
67 Ascott (1964) 42.
68 Kate Sloan Interview with Roy Ascott (2012) *Op. Cit*.
69 Ascott. *Op. Cit*. (1964) 40.
70 Shanken (2007) 205.
71 Townshend (2012) 256.

6 Game
Theatre | performance | play

The cybernetics of play

This chapter outlines the strains of game design and play that formed part of Groundcourse at Ealing and Ipswich, placing them in the context of a cybernetic theory of play and relating it to the dispersing aesthetic and conceptual influence of Cold War technologies. Games and play formed one of the most important elements of the Groundcourse curriculum. Play was a natural point of convergence for the many strands of the course in that it is social, behavioural and interactive. Game design itself is the act of creating a system for dynamic communication, and in this sense game design exercises directed the students to consider methods of controlling or manipulating a group situation. Within the course pedagogy, games served as exercises in systemic thinking, in codified design, in systems awareness and as tools for behavioural analysis. Curiously, the games design activities provoked a kind of self-conscious tension in Groundcourse participants; the knowledge of surveillance appeared to provoke a weighty, concentrated form of play. In 'performing' the games under conditions of observation, there were rich and surprising parallels with the secret and pressurized environment of the military bunker, bringing to life what Churchill would later describe as a 'secret war'.[1] In the symbolism and dramatic play of the games constructed by the students often evidenced what we might describe as a Cold War aesthetic: communications transmitted in confined, secretive environments.

Game play and its incarnations through time can tell us about social mores, traditions both aesthetic and cultural. They are engineered frameworks for human interaction, a matrix of social values that comes to life through play. There is an often unspoken relationship between the theatrics of play in art games and the aesthetics of cold war technologies, which relates to the Groundcourse games as a form of experimental cybernetic play. In game play, the innate interactivity of early cybernetic art reached a conclusive point, its various elements combining to form a communication system. This process took place in tandem with a broader trend in the visual arts towards using combinations of media as a layered and systematized approach. While the most prominent outcome of this trend was what would eventually become known as 'multimedia', multimedia is a limiting terminology for work that integrated many elements, both in terms of media and concept. Indeed, when in 1967 the artist Willoughby Sharp expressed the trend towards the use of combined media as 'total integration', he coined a phrase that better expressed the dissolution of elements that came about through combination, and the way such work existed through, between and because of combination.[2]

Certainly, Groundcourse play was not confined to a board and pieces; it was a theatrical strain of performance and set design that employed light, costume, sound and props and in this sense, it was multimedia. The experimental games produced by Ascott, his colleagues and his students worked on what was then a visionary precept – that the work of art existed in a web of relationships and exchanges, not as an object. The liquid and shifting performances of light, sound and play that took place at Groundcourse in the first half of the 1960s might have been produced with limited resources, but they nonetheless represented the same values of participation, feedback and performance as later cybernetic projects such as, for example, Experiments in Art and Technology's *9 Evenings* of 1966. In light of this, it is worth again reinforcing that as a country on the cusp of economic recovery after a long period of recession following the war, British artists were often more limited in terms of media and experimentation than those that benefitted from the burst of art–science collaborations in the US. In this chapter then, I draw out the innovative concepts informing Groundcourse play and its cybernetic inspiration, perhaps obscured by the meagre materials from which the games were constructed.

From radar to video games: military technologies of play

In the years that followed World War II, game play was to change beyond recognition as increasingly sophisticated new games were developed. This was not confined to the development of electronic gaming – other games that used traditional formats such as boards, cards and pieces also thrived. Much of this development related to the war and a brief overview of the origin of electronic gaming will allow us to draw out this peculiar relationship. The age of video games was born with the invention of radar, as the first game, designed by Thomas T. Goldsmith Junior in 1947, was a missile simulator.[3] It was based upon World War II and its circuitry was analogue. The only information displayed on the screen was a dot representing the 'missile' – the target and all other graphic information were overlays that were manually placed on the screen by the player, a layered experience of virtual and physical game play. The first digital computer designed specifically for playing a game was Ferranti's NIMROD computer. In 1952, it was exhibited as part of the Festival of Britain, and its display was formed from a panel of lights. The simple rules saw players move objects from groups, or 'heaps', with the player who removes the last object from the heap losing. The festival guide observed that 'It may appear that, in trying to make machines play games, we are wasting our time. This is not true as the theory of games is extremely complex and a machine that can play complex games can also be programmed to carry out complex practical problems.'[4] Despite the educative aims behind these game-display projects, the outcome was often quite simply, play. Stuckey, Swalwell and Ndalianis have observed that John Bennett, the software engineer behind NIMROD, was later disappointed that the audience wanted to '[...] play, rather than engage with the maths and science'. They also note that many of the young visitors to the festival retained nostalgia and fond memories of this proto-gaming experience.[5]

The most commonly recognized early electronic game *Pong* had as its non-commercial predecessor the game *Tennis for Two* which William Higinbotham developed in 1958 for the Brookhaven National Laboratory's annual visitor's day, under funding from the U.S. Department of Energy (Figures 6.1 and 6.2).[6] The display was intended to promote nuclear power and the computer itself was an analogue, run with an oscilloscope as a vector display system.

The installation photographs of *Tennis for Two* at BNL picture the game monitor under a sign reading 'computer tennis' that bore instructions. It was placed along a bank of computers with looping wires, switches and dials, under the large banner 'INSTRUMENTATION'. Another interface was labelled 'Electronic Counters'. It is evident that the functional element of these 'games' was all but obscured by the computers that generated them – computer tennis was a simple bouncing dot and line, moving on the ominously familiar thick glass of the radar screen. Oscilloscopes, like so many existing analogue technologies, had become more sophisticated during and immediately after World War II. Scientists had developed these analogue technologies while engineering radars and weapons. Early

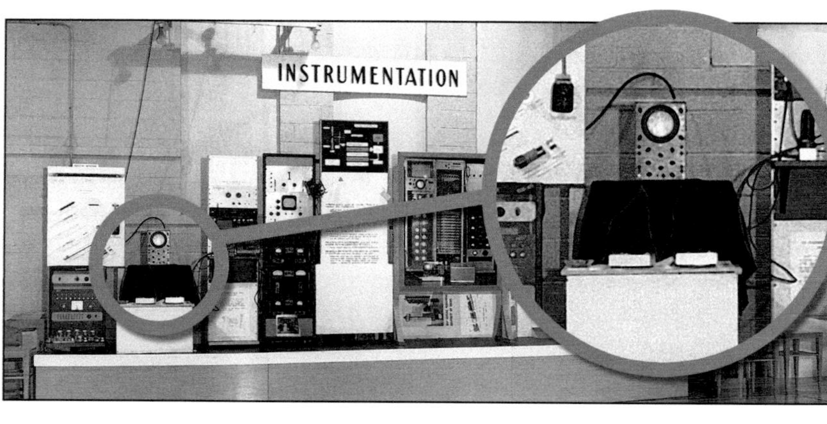

Figure 6.1 'Tennis for Two' on Display at the BNL Visitor Day. © Reproduced with the permission of Brookhaven National Laboratory

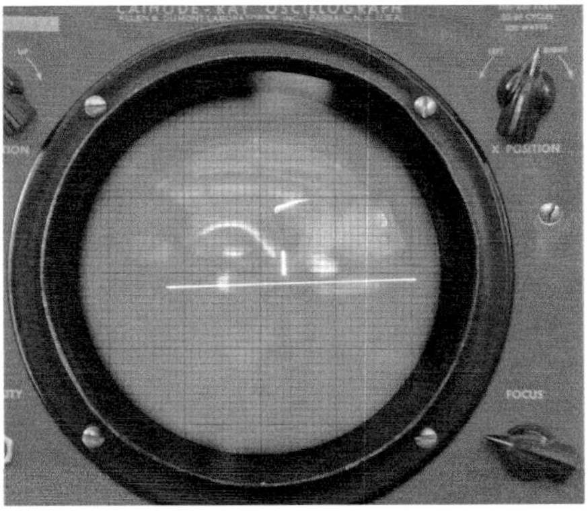

Figure 6.2 'Tennis for Two' on Display at the BNL Visitor Day. © Reproduced with the permission of Brookhaven National Laboratory

electronic games were the products of science labs in the post-war years, creating games as a method of public engagement. In this sense, game play was another attempt to overcome public distrust in technology. In the late 1950s and early 1960s, researchers used MIT's T-XO computer to create a number of interactive graphic displays, including *Tic-Tac-Toe*. Shannon, then based at MIT, developed an electronic *Mouse in the Maze* – a 'mouse' named Theseus who, like its namesake, had to navigate a maze. It learned complex routes through a 'maze' of twenty-five squares, seeking a goal that could be placed anywhere on the grid.[7]

The genesis of these early games was certainly prompted by the technologies of war, which were gradually moving from the realm of interactive technology education into play. In light of this there are a few points worth making with regards to the games above. Firstly, they were not developed with entertainment in mind and they were often display items used to promote science in an era of repressed fear. The reassuring tones applied during the Festival of Britain were part of an international movement – an attempted re-humanization of science in the wake of the dehumanizing and appalling destruction at Hiroshima and Nagasaki.[8] In the light of this, we must unpick this curious fact that the first known electronic game was based on missile trajectories – and developed a mere two years after the end of the war. The origin of the electronic game industry was never the pursuit of entertainment or play. It was a demonstration of cybernetic interaction, allowing an audience to understand what it meant to work with and through machines, in anticipation of the near future of machine dependency. The single pulsing 'missile' dot on the screen opened a world of projected possibilities, as the 'player' placed overlays over the screen and took aim. The act of playing this game was a simulation of war; yet it lacked the distance of simulation as we know it since it employed the same technologies used in war itself. It should also be noted that the fact these games were used in the context of festivals, open days and other public engagement fora indicates that they were intended not only to demystify technology but to educate through play. However, when the single pulsing dot on the radar screen transitioned from missile to tennis ball in these early exploratory games, electronic gaming began to move away from its origins.

While military-inspired forms of play had been in existence for centuries, the twentieth century saw two spikes in military-inspired games, films and other forms of entertainment; first after World War I and then after World War II. In the interwar years, the broad popularity of the game 'battleships' exemplifies this. The early electronic games discussed here were only one facet of the impact of World War II upon games – other gaming strategies such as the Groundcourse games used behavioural, systemic and controlled designs for play. At the same time, there was a growing interest in the sociology of play, resulting in an age of development for the nation's playgrounds and facilities, as well as a commercial explosion of new board games.[9] From this, we can take the fact that play was powerful; it could be used to channel certain forms of behaviour, both social or physical. It is also worth emphasizing that game design is in essence systems-based; it structures a group interaction and employs devices that can either disrupt or progress a player. There are, then, three layers to this confluence of war and play: games that recreated battles or weapons, games that repurposed war technologies and games that operated upon behavioural principles of interaction.

Game play doubled back at its point of origin and it became a military training or assessment strategy. The use of computer games and role-play games in the military occupies a peculiar space, given the popularity of leisure computer games based on war and other violent situations. There is an elastic line between the two given that they are only separated by intent; that is to say the gamer emulates war for the purposes of play and the army recruit uses play to practice, finally emulating behaviours learnt through play in combat. Given that in contemporary times, video games form an ordinary part of military training, play evidently takes in behavioural conditioning, performance and skills development, while all the time influencing popular forms of entertainment.

It is interesting then, that computer game play made its transition into light-hearted entertainment – from education to leisure – through the social interaction that is play. This did not happen instantaneously and the games discussed in this chapter were often serious, veering more towards a form of performance art than towards the pursuit of entertainment. The self-conscious physicality of early gaming has two layers – first, the echoes of war described here. In these early research 'games', pushing the red button held a complex dual implication of simulation and re-enactment. The second layer was the necessary interaction of physical objects and digital displays – the same digital-analogue tensions that formed the crux of the Groundcourse game. Just as the bunker operated around the curious tension of palpable, analogue objects and expanded virtual spaces, so did the game. This was equally true of the performance of technologies enacted by Groundcourse students and the first electronic games developed by scientists in the same time period.

For cyberneticists, machine gameplay served as the perfect illustration of the tangled philosophical problem of technology, agency and the possibility of artificial life. Throughout his career, Norbert Wiener was interested in the concept of human-machine gameplay, particularly with regards to existing games such as checkers and chess. He wrote speculatively of chess-playing machines in *Cybernetics: Or Control and Communication in the Animal and the Machine*, meditating that the problem was how to '[...] construct a machine which shall offer interesting opposition to a player at some one of the many levels at which human chess players find themselves'.[10] Later, while exploring cybernetics and religion in *God and Golem*, Wiener used machines that had the capacity to learn and adapt. He discusses successful checker playing machines and less able chess-playing machines.[11] Wiener directed his interest in cybernetic play towards the intellectual problem of the machine-player and how this might relate to human gameplay as well as to human responses to new technology. Alan Turing designed a chess programme in 1947, before there was a powerful enough computer to run it. Chess is the ultimate intellectual game and as such, it is little surprise that it served as an almost mythic goal for the artificial intelligence community.[12] It is also worth recalling Marcel Duchamp's observation that 'A game of chess is something plastic. You build it. It is a mechanical sculpture.'[13]

In Britain, Gordon Pask had developed a much more playful approach to human-machine interactions. Notably, his *Colloquy of Mobiles* for the 1968 *Cybernetic Serendipity* exhibition at the ICA, which employed light-responsive techniques to create a participatory interaction between the viewers and the mobile. The visitors shone torches at the mobiles, triggering mechanical responses that Pask had modelled as analogical to a male/female reproductive ritual. Pask had also created a sound-responsive theatre light show entitled *Musicolour* that performed in 1953 with limited success. Collaborating with Robert McKinnon-Wood, Pask was exploring an '[...] interest in synesthesia and the question of whether a machine could learn relations between sounds and visual patterns and in doing so enhance a musical performance'.[14] *Musicolour* was certainly a form of play – Pask and McKinnon-Wood had designed it so that the musician would 'learn' how the system responded to their music and if they liked certain light values, they would reinforce passages of the music. The system in turn would respond to this reinforcement. However, *Musicolour* was also configured to adjust itself if there was too much repetition in the musical inputs, creating more variety.[15] As Wood and Di Paolo have expressed it:

> From the outset, Musicolour was designed to cooperate with human performers, rather than autonomously generate 'aesthetically valuable output' [...] The way musicians interacted with the system quickly

> became the main focus of research and development: the performer 'trained the machine and it played a game with him'.[16]

The Paskian model of play was intuitive and, vitally, it was responsive in both directions. Both man and machine adapted because both were part of the same system, creating sound and light that fused to make a single performance. Imperfect, marred by unsuccessful attempts to use it alongside puppetry in performance and less impressive in performance than in theory, *Musicolour* nonetheless embodied cybernetic play in an arrestingly eccentric form.

While the art produced during Groundcourse was not engineered, it operated on the explorative, systemic and communicative territory that was the human face of cybernetics. Bearing in mind the strong visual imprint of the cipher-breaker manifested in the student's analogue sculpture, Ascott's rhetoric holds arresting possibilities for the deconstruction of the games that resulted from this exercise. Creating and breaking code was the greatest military operation of World War II and remained a critical focus in the Cold War. We must question here whether code breaking can simply be transferred to a detached form of artistic 'play', when the dual meaning of code breaking as a military operation was such a prevalent facet of post-war culture. Given that the late twentieth century saw the birth of the gaming industry, the moment in time described here becomes anticipatory, prescient even. While we know that games imitate or reimagine the real world, there is also further significance of Ascott's time in ground control in relation to game design. Since the post-war age of communication technologies began, gaming has become wired in, connective, virtual, analogous and digital.

Ascott's pedagogy and play

Ascott's interest in play was embedded in his larger preoccupation with interaction and the transmission of ideas. In 'The Construction of Change', Ascott framed interactivity as 'behavioural', a formal way to describe what was often a playful encounter. He described participant interaction with a work of art in a ludic way:

> In response to behavioural clues in a construction (to push, pull, slide back, open, peg-in, for example) the participant becomes responsible for the extension of the artwork's meaning. He becomes a decision-maker in the symbolic world which confronts him.[17]

Ascott's analogues and games all worked on the premise that physical elements such as rods, pegs, panels or levers would be manipulated by the viewer, transforming the viewer to participant in a performance of the work of art. It is also true that this form of abstract interactivity relies on the same conceptual understanding as a board game, in which a defined set of choices result in the physical manipulation of pieces or markers, creating paths and routes through the game. In each case, a set of analogical relationships exists between the markers and some kind of contextual meaning, whether it is aesthetic or metaphoric. By describing this process as behavioural, Ascott emphasized that interaction was agency. In this light, it is unsurprising that play emerged as Ascott's dominant strategy in art and pedagogy in the 1960s. Play encompassed both Ascott's precept that art was 'creating forms and structures that embody' ideas, and his suggestion that as spectators altered the 'formal relationships', so the ideas these formal relationships represent are extended or changed.[18] It does not take much of a leap to recontextualize Ascott's analogues not as interactive sculpture but as matrices of play, frameworks of interaction that were designed to make the viewer-participant aware of their own place in the generation of meaning. Tim Stott's term 'ludic participation' is a helpful description of this strain of participatory playfulness in the visual arts.[19] It is notable that post-war forms of concentrated play – in art, cybernetics, child development, anthropology and the social sciences – operated upon a definitively analogical model. That is to say, play was not aimless; it always connected (directly or indirectly) to some other social or cultural quantity. In this sense, ludic participation in the time of Groundcourse was part of an emerging phenomenon, with a distinctly systematized conception of play, far from the chaos-seeking models of Dada. Claire Bishop has written of the need for a new approach to participatory play projects in modern and contemporary art:

> Just as we have come to recognise Dada cabaret, Situationist *detournement*, or dematerialised conceptual and performance art as having their own aesthetics of production and circulation, so too do the often formless-looking photo-documents of participatory projects have their own experiential regime.[20]

Certainly, the arresting records of Groundcourse games fit into an amorphous category of documentary records of 1960s art projects, more often described as 'happenings'. In the context of art school pedagogy, the relative value of modern archival records was often overlooked, leading to a loss of archival material.[21] This was particularly problematic during the period of academicization of fine art in the UK, when many colleges were combined into new polytechnics or merged into Universities.[22] It may be useful to reconfigure approaches to those organized interactive events of the early 1960s that involved rules, tasks or processes and to place them firmly within a framework of ludic participation.

As Ascott developed his earliest pedagogy, he often used ideas that he extracted and distilled from his own practice. This reflected the 'laboratory' ethos that Pasmore had applied at King's College and it was also in keeping with a broader trend in UK art schools towards a quasi-collaborative approach.[23] The tasks of calibration and mind-mapping detailed in the previous chapter were designed to give the students a sense of the freedom of an open field of communication. In this sense, when Ascott wrote 'The Construction of Change' he was articulating the end of the art object as a defined and stable entity, and proposing a new attitude to making art, in which you were constructing open-ended possibility. The two groups of Groundcourse students (at Ealing and then at Ipswich) were each placed in a situation that fostered playfulness; they worked with light, environment, costume, setting, interactivity, game design and performance. Taken together with the core principles of interdependence and group work, the course quite naturally evolved over its two incarnations towards a format that resembled ongoing play.

Codes, grids, matrices

Groundcourse students were explicitly tasked with game design at several points in the course. This included directly designing the visual material for a game, its structure and its rules. Because gameplay has rules and outcomes not just actions, the student-artist had to create the framework of limitations. Ascott describes this as a process in which 'Students set about analysing and inventing games, logical propositions, idea sequences, and matrices. Visual polemic is induced, and codes are designed and broken.'[24] This was essentially an experiment in possibility – how related visual forms, diagrams or instructions created or limited choices and how the outcomes of play could be defined or controlled. The work that students produced during these exercises often exhibited a formal and repetitive quality as grids opened out to cubes, cubes to matrices, matrices to boundless possibility (Figures 6.3 and 6.4). This was an extraordinary language of sequences and matrices, propositions, polemic and codes, manifested in limited contrasting colours and repeating, mutating structures.

The top image shows a wall of drawings of complex circuits and repeated patterns, as well as an analogue machine with a crank handle and wiring sitting on the floor before them. The analytical quality of these diagrammatic drawings is striking; in black, white and block colour, they resemble the working drawings of engineers. Observing these images from the distance of a new century, they resemble the technologies of the age caught in time: wall of circuits, of codified colour, of complex, interconnected units. The sequential drawings reflect the complexity of the game designs, in that they were logical rather than intuitive, often mathematical, always ordered. These drawings sit in almost linear contrast to the Dada and Surrealist forms of game play that followed World War I. Systemic and positivist, they start from a principle of controlled design that stands at odds with often nihilistic or fruitless interwar forms of play that focused on chance rather than design. The balance of these twin poles of game play defines both the mode of play and the range of potential outcomes. Games such as *Exquisite Corpse* or Tzara's Surrealist poems had courted randomness and unpredictability; they were systems for creating art by chance. The Groundcourse games, in contrast, visually described the principles of behaviourism, systems of control, cybernetic communication and technology that informed the course.

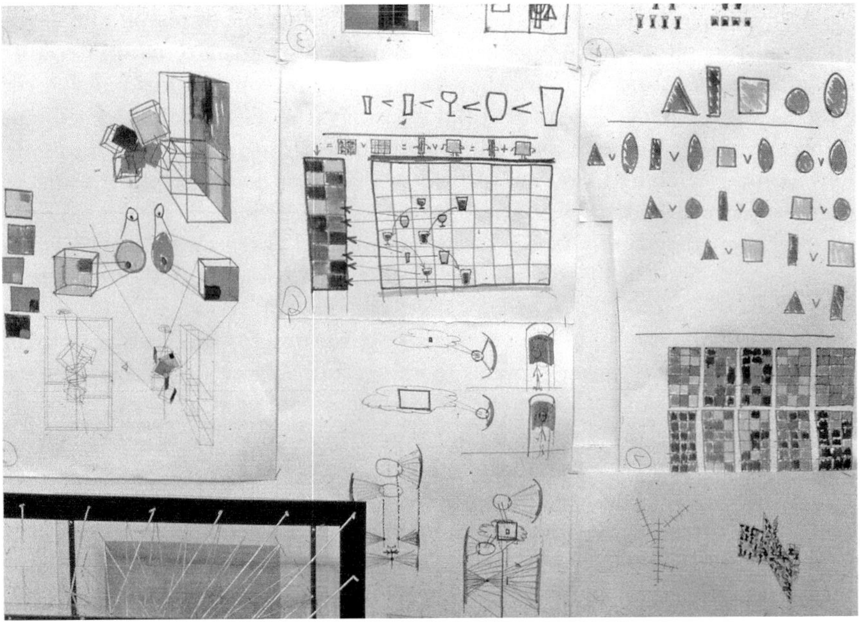

Figure 6.3 Roy Ascott (c. 1964) *Student Drawings Showing Games, Analogues and Systems*. Photograph © Roy Ascott

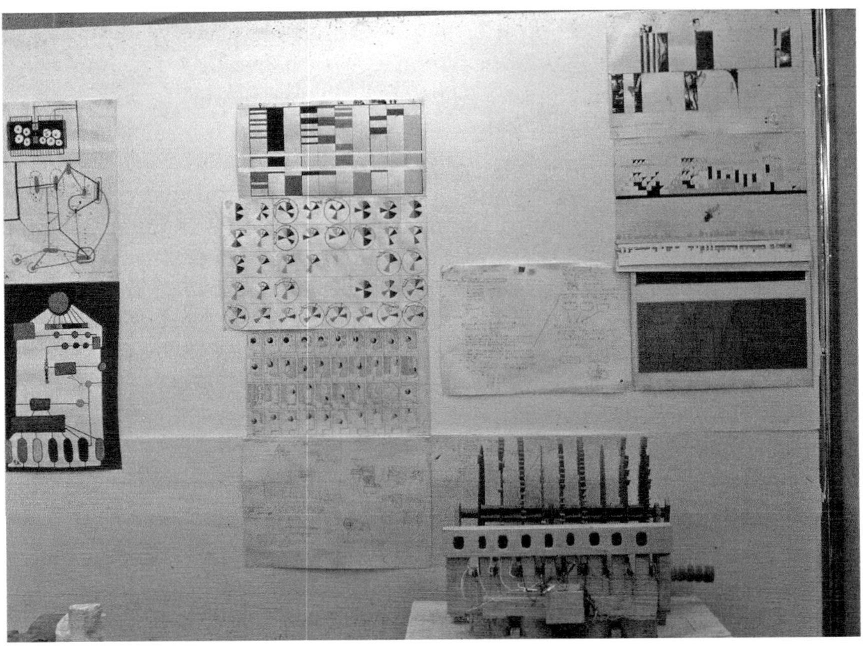

Figure 6.4 Roy Ascott (c. 1964) *Student Drawings Showing Games, Analogues and Systems*. Photograph © Roy Ascott

These drawings have a common codified quality, with shape and form mutating into sequence. As geometric shapes, vessels, patterns and colour sequences repeat and change, they bring to mind the evolutionary form sequences in D'Arcy Wentworth Thompson's *On Growth and Form*, in which a single form such as a fish or an animal skull would be gradually skewed on a stretched or slanted grid in order to demonstrate the mathematics of living structures.[25] There were only a few years between Ascott's immersion in the King's College curriculum and the creation of his own curriculum design for Groundcourse. Ascott used some ideas that had their roots in *On Growth and Form*, particularly that of evolving form, in the Groundcourse curriculum. However, these game designs used evolving form as a functional code rather than as an end result. Ascott's form had behaviour, and during the two incarnations of Groundcourse, provoking, controlling, manipulating or reacting to behaviour became the core approach to visual arts production for staff and students.

The drawing in the centre of Figure 6.3 was evidently a design for a game based around different shapes and sizes of drinking glasses on a chequerboard. The game markers are connected but distinct. Other drawings on the same wall show different grid colour formations, series of geometric shapes in primary colours, sequences of symbols evolving in shape and colour and complex connected grids like electrical circuits. The process of invention with a game involves an analogue mode of creative practice, developing a set of possibilities and a way to visually represent them. Making the rules and setting the boundaries is a practice of design that is ingrained with the behaviourist approach to provoking and analyzing certain modes of human response. The grid, the map and the route formed the backdrop to the manipulation of these signs and symbols, just as they do in the board game; in this sense the matrix was the core of the game.

A matrix operated as an organizational structure or grid that could physically measure progression through stages of play, while shaping the concept of the game. The grid designs on some of the drawings to the left of Figure 6.4 have a nodal quality, networks of components joined by wires. The students were familiar with the concept of connectivity, particularly in relation to their own roles as part of the 'organism' of the course. These drawings are similar to the mind maps also produced at Ealing which had traced relationships between actions, possibilities and decisions.[26] In a sense then, these were network concepts realized in visual form. Ascott had a holistic belief in the universality of systems behaviour, which chimes with the curious and harmonious cross-disciplinary General Systems Theory that Ludwig von Bertalanffy would describe in 1969:

> [...] there exist models, principles, and laws that apply to generalised systems or their subclasses, irrespective of their particular kind, the nature of their component elements, and the relationships or 'forces' between them. It seems legitimate to ask for a theory, not of systems

> of a more or less special kind, but of universal principles applying to systems in general.[27]

Von Bertalanffy's theorized systemic behaviour in a way that he hoped would be helpful across disciplines, anticipating the need to manage systems to maximize productivity and output, to understand growth and development, and to understand the ramifications of change within a complex network of interrelated human and mechanical elements. During the war, General Systems Theory had evolved from the interrelated logistics of complex elements including intelligence, the movement of soldiers, food, weapons and water, and von Bertalanffy was the most important contributor to the field. He described the cybernetic convergences of wartime engineering:

> [...] when it comes to ballistic missiles or space vehicles, they have to be assembled from components originating in heterogeneous technologies, mechanical, electronic, chemical etc.; relations of man and machine come into play; and innumerable financial, economic, social and political problems are thrown into the bargain.[28]

For von Bertalanffy, establishing a General Systems Theory was a means to understanding and managing the complex interrelationships that had also been the basis of the field of cybernetics. The complex communications and interactions between machines and people were, as he saw it, indicative of a new era for the sciences. In a famous passage of writing he claims that 'The 19th and first half of the 20th century conceived of the *world as chaos* [...] Now we are looking for another basic outlook on the world: *the world as organization*.'[29] The world 'as organization' is certainly a compelling and poetic theoretical model for the post-war growth in cybernetics, behaviourism, communication technologies and organizational management. A game board, an electric circuit or a social network diagram share similar qualities once depicted visually; that is, they are systems built of hubs and relationships. For von Bertalanffy, treating *the world as organization* had the potential to '[...] help to reinforce the sense of reverence for the living which we have almost lost in the last sanguinary decades of human history'.[30] In the immediate post-war years, the systems developments that had been provoked by the war were reconceptualized by a generation of scientists and engineers as positive tools for recovery and advancement. In this sense, it is easier to understand why Ascott and some of his contemporaries saw cybernetics and associated communication technologies as visions of an enlightened future.

Some of the most arresting photographs of the Groundcourse games records students playing in a 3D matrix (Figure 6.5). The students constructed a grid which became a three-dimensional game. The players negotiate it, climb from it, hang from it. The grid is the matrix of the game and the players in this context move physically through the matrix itself. This

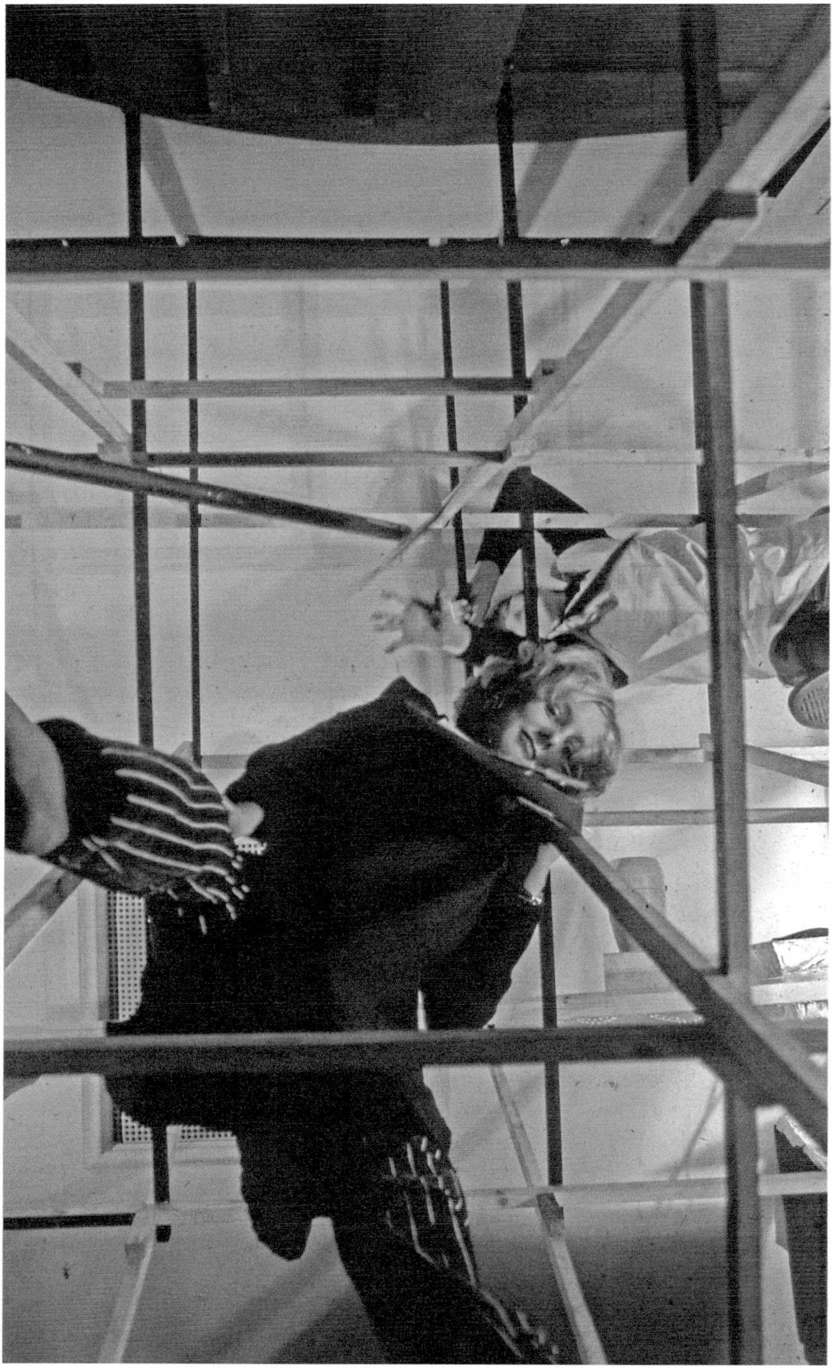

Figure 6.5 Roy Ascott (1965) *Groundcourse Behavioural Project.* Ipswich. Photograph © Roy Ascott

physical encounter extended the game play into an environmental experience, exploring the dynamics of this grid form which played upon the logic of game design and opened it out into the immediate environment.

This game/set has echoes of the utilitarian structures of underground bunkers, which were functionally constructed and often had exposed beams, supports and pipes. The eerily lit structure retains a sense of utilitarian architecture, of functionality, despite being used for play. It is worth noting that grid structures were used by several artists and designers of the period, as well as their widespread application in architecture, particularly Brutalist projects. Hamilton employed grid structures as an exhibition display device several times during the 1950s and 1960s, and the suggestive logic and mathematical harmony of the grid indicated reason, logic and order.[31] In this context, however, a three-dimensional grid became a matrix, an open set of possibilities for play. It is evident then that Groundcourse games moved swiftly beyond the concept of a board or set, opening out into real space and intensely focused upon the players and their behaviour.

Bunker aesthetics: code in colour

I have already discussed Ascott's experience of Fighter Control as an extended or analogic environment.[32] There is another relevant and often-overlooked element to Ascott's bunker experience and that is its vibrant, surprising and game-like aesthetic. In the bunker, cool painted grey metal offset the bright primary colours employed in objects including ops clocks, plane markers, maps and telephones. The same offset of primary colour against neutrals was also prevalent in post-war game design, including the work produced by Groundcourse students. The matrix drawings employed red, yellow and blue with black, white and grey, a palette that has long been employed in game design but that had recently become essential to an increasingly coded military environment in which streams of information and multiple functions had to be clearly denoted with maximum contrast. The fact that images of World War II exist in black and white means that this chromatic vibrancy is little-discussed, despite the fact it is so arrestingly connected to Cold War aesthetics in both art and game design. In game design, the aesthetics of war technologies contorted, re-emerging as play. In the Cold War bunker, coloured plastic, metal and wooden pieces were the moving elements within a context of analogous display. Curiously then, the cybernetic environment of the bunker involved the same visual language as the board game, from the moving pieces to the coded language of symbol, colour and shape. Consider for example, the distinctive environment of the operations room at RAF Uxbridge (Figures 6.6 and 6.7) from which the Battle of Britain was coordinated. The table-top map and its pieces were manipulated by a large team, each with a long wooden sweeper and each plugged into a set of headphones.

The image of the empty table reduces the ops room to a set with its props, and in so doing makes it easy to deduce the similarities to board games. The

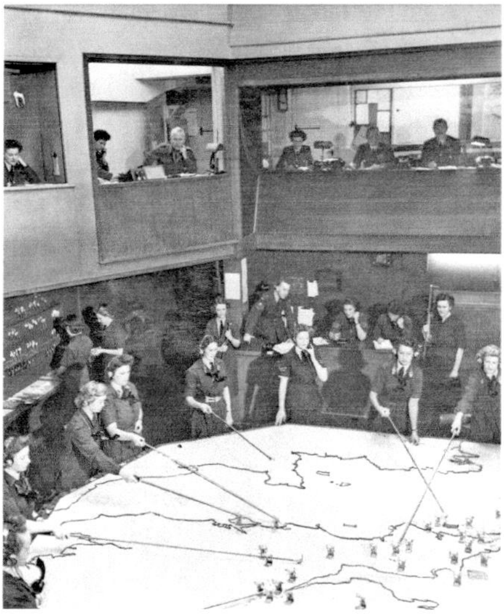

Figure 6.6 Operations Room, Battle of Britain. Photograph © Imperial War Museum

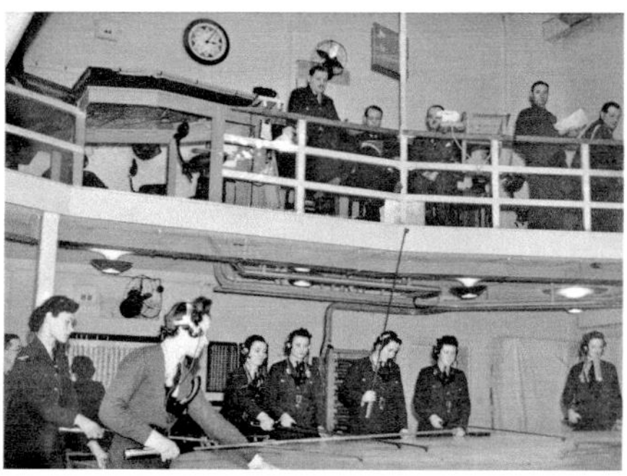

Figure 6.7 Operations Room, the Battle of Britain. Photograph © Imperial War Museum

wooden sweepers and pieces wait for the players to advance, and the map on the table surface could so easily be the matrix of play. Each of the operatives surrounding the board wears a headset from which they receive the information they need from colleagues at the radar screens – they are physically wired in. In this sense, the ops room formed a cybernetic system in which information was exchanged virtually, and represented analogically. The boards and set-ups for the Groundcourse games have clear visual links to the cybernetic ground control environment too. The hexagonal board below (Plate XI) draws on circuitry, its base the gunmetal colour that was used on early computer monitors, on circuit boards and on the bases of hundreds of mechanical elements within the bunker environment. It is a colour that should not be considered neutral.

Ascott recalls that within the environment of ground control there were pentagonal screens. He also mused that while group work had been set to groups of six, he wished it had been five, since it was a better number.[33] Most Groundcourse group activities had been performed in groups of six and the hexagonal form of this board perhaps echoes this. Each shape precisely subdivides into triangular segments correlating to its number of sides, and in this sense the pentagon and the hexagon are good illustrations of a collective of individuals. On the board, underneath the black circles are a series of geometric lines in red and black, forming a grid of parallelograms and triangles. There is an element of sci-fi abstraction behind this game design; something of the dreamed-up spaceship or a sophisticated James Bond villain's lair in which atomic war is planned and foiled. The technological aesthetic of this game board also collides with the traditional; the wooden base, with its incised lines and bright primary triangles and circles, brings to mind a Backgammon board. The language of games evolved over centuries into a language of primary colours, grids, neutrals and blood red.

The same coded language informed military environments including bunkers, the insides of military vehicles and the exteriors of computers and weapons. In the existing black and white photographs of Fighter Control bunkers and operations rooms, the intense colours that were employed in the design of military technology are rendered invisible. A mantle of grey has fallen over our technological history – the colour of a computer case, an interface board or a control panel. However, grey formed the backdrop to a far more distinctive palette of design, in which vital function was indicated with the brightest colours. In both Fighter Control and the Groundcourse games, bright, primary colours and black and white were used to signify function and meaning. We might remember that pressing the red button has become a metaphor for the forbidden technological act; it will send the missile into the air, trigger the blast or send the ejection seat soaring. In the world of machines, bright colour indicates urgency, agency and at times, danger. Much like the jewel-coloured casings of poisonous insects, the primary colours of military technologies are a warning.

This application of colour is purely code – the natural contrast of primary colours meant that visual differentiation would be simplified in what was a rapidly changing environment. The most important symbolic function of colour in Fighter Control was its application in the ops clock – the triangles of primary colour formed five-minute time intervals so that operatives could colour code their markers according to when a reading was taken (see Plates VIII & IX). The triangulated position of the marker upon the map formed the vital information which would be given to pilots as they flew to intercept potential enemies. Each control room contained an ops clock during World War II. In the frantic environment of the ops room, when each operator received information about a raider crossing into British waters, they would colour code their marker within the five-minute interval on the clock. This information was crucial as it had to be accurate to within these time limits in order for fighters in the air to make their interception.

Primary colours are ideal for coding given that they are by nature the most starkly different from each other in their unblended state. Black, white and grey, used for backgrounds, are neutral. Black and white grids, structures and diagrams form a matrix for information and the coloured elements are the variables. They are also the most reductive palette for any game: black and white the matrix of possibility and blood the human element, the element of life. The attack warning telephones, such as those used in RAF Neatishead (Figure 6.8), are bright red. This former Royal Observer Corps Station is now preserved as a museum.

The colour scheme of the attack warning phone, of neutral grey with red, black and white, has a strong visual link to the scheme used for the board game in Figure 6.9. The student work created during Groundcourse had a distinctive and recognizable palette. A palette of primary colours, black, white and gunmetal grey. It was the coded colour scheme of war and its technologies. In the Fighter Control bunker, the red telephone stood for emergency; if Ascott or his colleagues identified the potential threat of an unidentified plane, they would seize a red telephone and utter the words 'Scramble! Scramble! Scramble!' to the operator. This would send a jet straight out over the sea to investigate. In the late 1960s, when Ascott was the president of the Ontario College of Art, he created a 'power house'

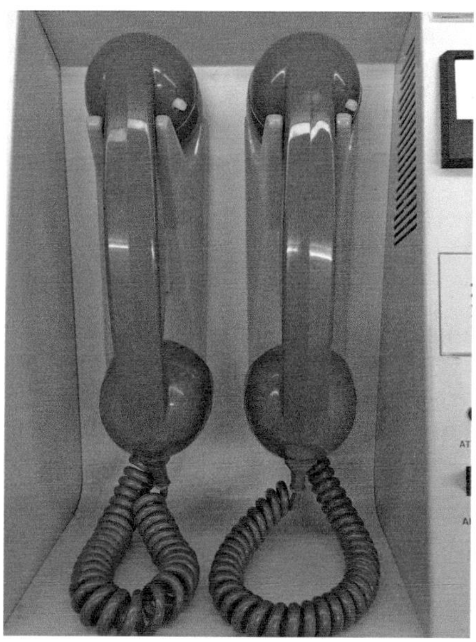

Figure 6.8 Attack Warning telephone. RAF Air Defence Radar Museum, Neatishead, Norfolk

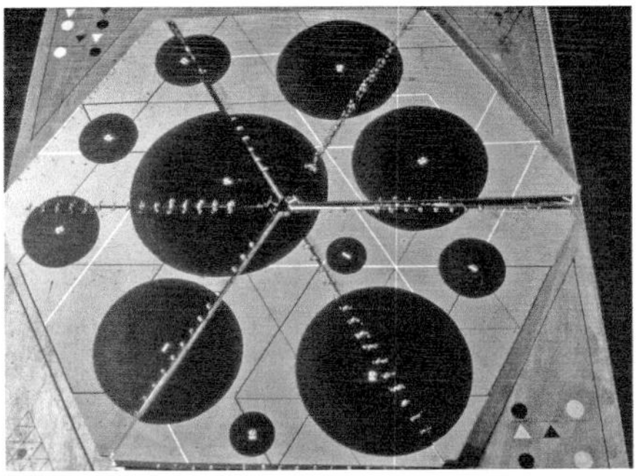

Figure 6.9 Roy Ascott (1965) *Groundcourse Board Game*. Photograph © Roy Ascott

set-up for teaching staff that drew on the ops room urgency. Teaching staff were seated at a round table with dangling wires and telephones, through which students would urgently communicate ideas.[34] While this radical experiment did not last, it certainly shows that for Ascott, the telephone still symbolized urgency, information exchange and action.

Performing surveillance and the sociology of play

Game play relies on artifice: costume, props and preconceived formats of play, including markers or pieces that symbolically represent the player, a value or an idea. The Groundcourse game design exercises were created in full recognition of the fact play was a social construction. In a Groundcourse curriculum document, tutor Lawrence Self set the parameters for a game design project at Ipswich. In the page-long hand-out, Self offered an interpretation of game design that almost verged on set design:

> This would necessitate the creation of an environment within a determined and separate area. The structure would require a variety of materials. Light, colour, sound, scent and texture, together with movement and change could be considered as possible attributes. The function of the group in relation to this environment might require the use of masks, costumes, voices, music as the game develops.[35]

In this description of play, Self outlined a model of game design as an environment of play, in which all sensations might be manipulated as part of play. There are two layers to the description above – the first layer is like a multimedia stage set in which the manipulation of light, sound or scent are treated as 'materials'. The second is artifice – masks, costume and the modulation of voices are performative and potentially misleading. Self presented game play as an elaborate and testing experience, one which the students were to design and direct.

Self quoted extensively from the 1961 sociological exploration of game play by Roger Caillois, *Man, Play and Games*, listing in full the attitudes Caillois ascribed to the 'marginal and abstract world of play':

> The desire to challenge, and make a record, or morally overcome an obstacle
> The hope for and the pursuit of the favour of destiny
> Pleasure in secrecy, make-believe, or disguise
> Fear or inspiring of fear
> The search for repetition and symmetry, or in contrast, the joy of improvising, inventing or indefinitely varying solutions
> Solving a mystery or riddle
> The satisfaction from all arts involving contrivance
> The desire to test one's strength, skill, speed, endurance, equilibrium, or ingenuity
> Conformity to rules and laws, the duty to respect them, and the temptation to circumvent them
> And lastly, the intoxication, longing for ecstasy and the desire for voluptuous panic[36]

This extraordinary list operates around the tension of living and acting, condensing play into a series of sensations, including fear, triumph, competition, panic, joy and hope. Caillois positioned game play as a stage for our human impulses and desires. In setting this exercise, Self encouraged students to analyze game play as a social structure, to pick out relationships, interactions and transactions.

It is also worth picking out some of the elements of this list that certainly influenced the Ipswich games, notably disguise, secrecy and fear. Since the principle of the game as sociological model was the starting point for the exercise, it is notable that the formalizing element of surveillance became an integral facet of play. Surveillance creates a power dynamic between player and observer. The player has agency but the observer can analyze, the player can create but the observer can judge. In conventional game play, the act of observing play is made democratic by the fact that everyone is playing and everyone is watched by their fellow players. In the Groundcourse games, surveillance took the form of an additional layer of observation, such as one group watching another play, or players who had 'lost' a game watching the other players from the wings. Just as observed game play had been used as a tool during the calibration and mind-mapping process, so it recurred in other exercises as a circuit of play, performance and analysis. The games present an interesting problem with regards to self-analysis, in that it is hard to respond naturally when you are both self-conscious and under observation. The Groundcourse notion of a counter-response – something contrary to your 'natural' course of action – essentially provided a framework for communication studies within the course. Students formed synthetic models to assess self-knowledge, and in employing these models, they slipped into performance.

If we view the Groundcourse games alongside other pedagogical experiments of the 1960s such as Kardia's *Locked Room* with sculpture staff and students involved in the Central St Martins 'A-Course', it is evident that a dynamic of performance and observation was manifest. At St Martin's, the *Locked Room* project was certainly built upon power and limitation. As Elizabeth Wright has noted: 'The staff themselves orchestrated the performance; as gatekeepers, the students had to negotiate if they could exit or enter outside of these fixed times.'[37] This is a curious set-up of willing imprisonment; as Hester Westley has described:

> Kardia established a pedagogy based on paradox: literally imprisoning students in order to liberate them. Garth Evans emphasised the importance of these negating practices: 'If one is not careful as an art teacher, one can believe that one has the obligation to tell a student what to do. What I learned in The Locked Room is that all one can intelligently tell a student is what not to do.'[38]

Experimental pedagogies that employed power tactics such as the *Locked Room* and Groundcourse did so in the spirit of exploring them. That is, the issue of relative power simultaneously generated a contemporary strain of visual research and a subversive pedagogical model.

In the 1960s, there was a very particular art school trend of projects that forced surreal role play such as those that were part of the Hammersmith College foundation course.[39] It was a phenomenon that arose from a Cold War culture of secrecy. In an era that saw warfare retreat underground, in which military secrecy became a political issue, in which fear and tension rested unhappily with social and economic recovery, the enacted tension of these art school experiments was certainly a product of an age. The students were subject to constant analysis – from self-analysis to parodied exercises in which staff observed students. In the games, the costumed, masked figures indicate student awareness of, and engagement with, the same tension of surveillance with which they had been tested and confronted throughout the course. In Figures 6.10 and 6.11 the player picks up a circular disk with deliberation, his head turned towards the player at his shoulder whose face is obscured.

In this parody of surveillance, we see both the impact of Cold War military tension upon the students of Groundcourse, and the impact of the integral behaviourism of the curriculum, culminating in a performance of ritualized tension. Bateson had described gameplay in quasi-ritualistic terms, writing of:

> [...] the dim region where art, magic, and religion meet and overlap, human beings have evolved the 'metaphor that is meant', the flag which men will die to save, and the sacrament that is felt to be more than 'an outward and visible sign, given unto us'.[40]

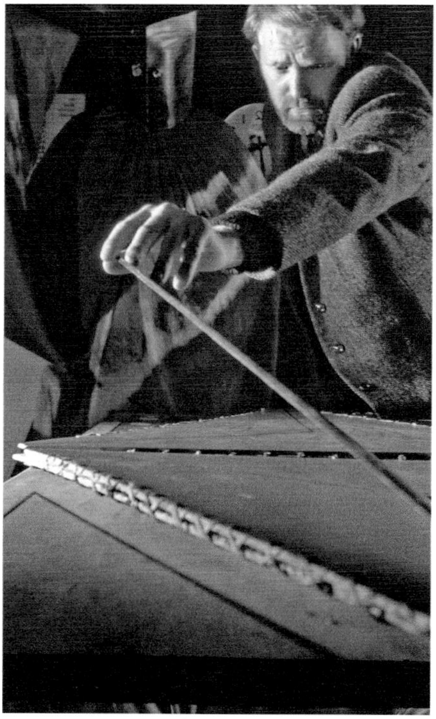

Figure 6.10 Roy Beston (1965) *Groundcourse Game, Ipswich*. Photograph courtesy of Roy Ascott

Bateson's evocative words reflect the recentness of the war. He conjured a very contemporary illustration of the power of a symbol: the flag, with its ability to impel loyalty even to death. Bateson explores the fact that we '[...] half consciously choose to blur the distinction between the information, signal or code and the real object or event'.[41] These game-performances exist in that realm between real events and fictions, blurring the symbols into a mysterious atmosphere of tension.

Self's project document framed the Ipswich Groundcourse game in a very particular way, urging the students towards forms of play that were both theatrical and behavioural. Some of the most striking surviving images of Groundcourse pedagogy are the photographs of the games played at Ipswich in 1965. The photographic record of these games demonstrates intense concentration on the part of the players, as well as self-consciousness in front of the lens. There was no boundary of ownership to these games because they were built, performed, watched and documented in groups. Staff and students participated in games that employed boards, set, costumes, lighting and physical movement. In some of the images, the hexagonal game board discussed earlier in this chapter (Figure 6.9 and Plate XI) is caught in use. A figure stretches (Plate X) across a board consisting of coloured hexagons

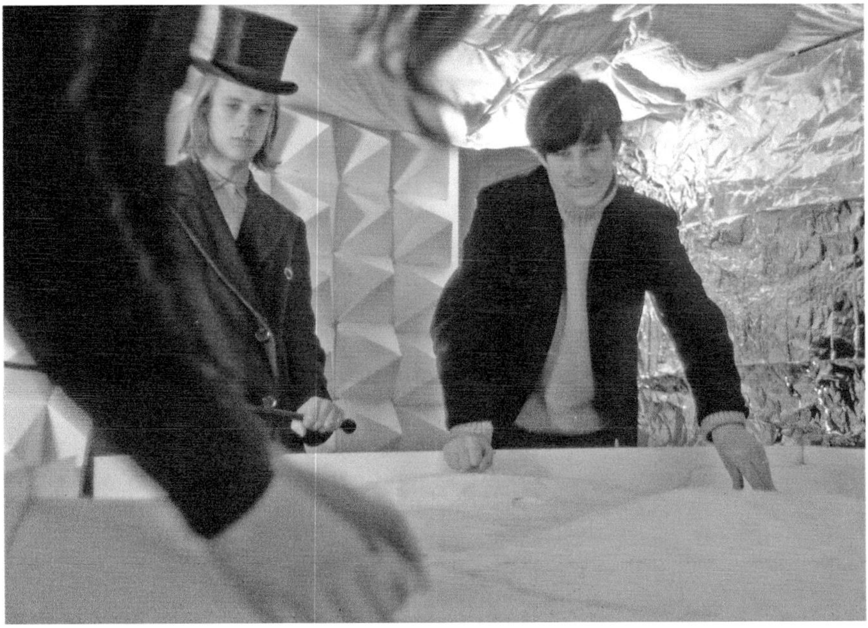

Figure 6.11 Roy Beston (1965) *Groundcourse Game, Ipswich. Brian Eno and two other players*. Photograph courtesy of Roy Ascott

made up of triangular 'flag' shapes that again recall the operations room and its palette of intense colours set against grey. This hexagonal board (Figure 6.9) unfolded outwards, reveals connected forms which the player had to stretch to touch.

In this game the body became a physical marker, positioned in relation to the coloured grid.

Shanken described this game as a precursor to the board game 'twister', with regards to the physical movements of the players.[42] The physicality of the game and the contortions of the player's body are quite evident. This formalized strain of play emphasized the phenomenological encounter at the heart of games; their physicality, enactment as a series of physical movements or gestures that operates on the boundary of the real and the imagined. In the photograph of this game, the player leans forward to position a long stick, apparently completely absorbed. In the background to the left, a dark figure is just visible, cloaked and concealed with a box or bag over the head with rough eyeholes. Ascott recalls that concealing headdresses and other forms of disguise were employed during this game when the player lost – or 'died'.[43] The staged malevolence of the photograph is no accident then; this game was an analogy for 'survival', with the losers 'dying'.

The player pictured here leans over the board manipulating a long stick, echoing how the RAF Fighter Control staff leaned over their tables with sweepers and pushed their markers into place (Figures 6.6 and 6.7). It is also notable that the lighting in the room – for this game and for others – was dramatic, at times flooded, at others subdued and shadowy, but never natural. The games are played out as if on a stage; the players could be in an underground bunker rather than in a college art studio. This dark and oppressive atmosphere is intensified by the presence of cloaked and masked figures within the game. In this curious performance there is a resonant echo of ground control, amplified by the cloaked atmosphere of surveillance. Certainly, the games themselves have a strange energy about them; many of the photographs show contortions or frozen poses, physical manipulations through play. There is a kind of tension about this conscious performance of 'the game' as a model for art practice. A printed set of instructions for a board game (Figure 6.12) shows the stark instructions to one student's game, black letters on red.

This game ran against the convention of scoring highly; the winner must score as little as possible as quickly as possible. The rules state that the 'last deity records scores', meaning that the last player, who has 'lived' through the game and 'died' at its completion, took up the role of the deity. The 'dead' players were given a tribal mask to wear, and it was a monstrous heavy, hairy disguise. The 'dead' lurked in the background, sacrificing their own identity to a presence of ritualized menace. At base level, this dramatic mode of play simply draws out the natural alignment of game play to living: a start point, a journey of progressions and losses and a final conclusion, or 'death'. However, the malevolence of the costumed presence of death, masked or

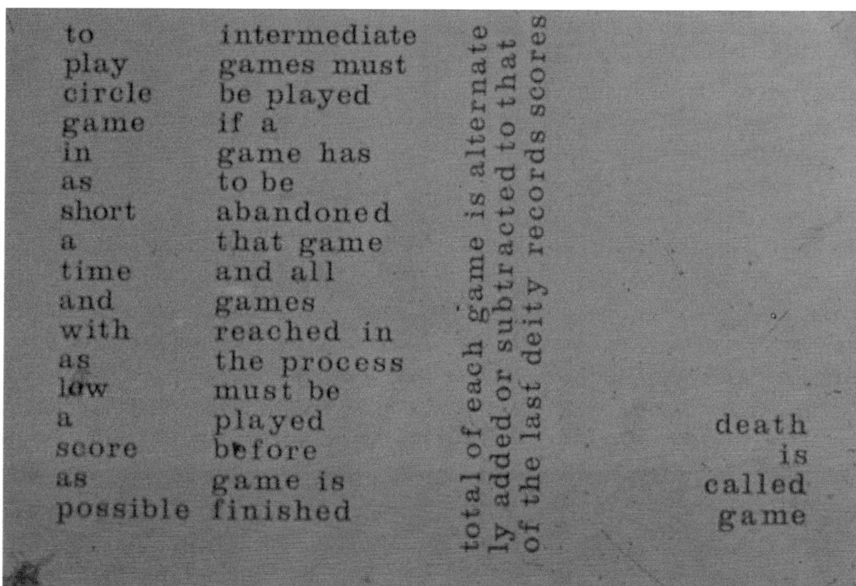

Figure 6.12 Roy Beston (1965) *Instructions for a Student Game. Ipswich.* Photograph courtesy of Roy Ascott

cloaked horribly in a bag with rough eyeholes, plays upon both movie horror and an awkward, undefined sense of 'otherness'. In his project curriculum document, Lawrence Self quoted the translator's introduction to Caillois' *Man, Play and Games*, in which he had mentioned the game play traditions in several other cultures, including Australian, African, American Indian, and ancient China and Rome.[44] It is quite probable that the elements of the masks and costumes reflect these literary and conceptual touchpoints that were provided by Self.

Self had also recommended a small number of books to the students, which included Karl Groos' *The Play of Man* (1901) and *The Play of Animals* (1898).[45] The introduction to *The Play of Man* is entitled 'The System of Play', which the author then attempts to create through the analysis of child's play first in relation to the senses and 'higher mental powers' and then in relation to the various forms play takes (fighting play, love play, imitative play, social play).[46] Groos' book configured play in relation to child development (running, throwing, catching), and then to the socialization of children as they learn about affection, collaboration and conflict. Groos discusses the natural way that a group of children will submit to the power of a ringleader, which he reads as evidence that '[...] play is instrumental in teaching children submission to law as well as to a leader'.[47] His earlier volume *The Play of Animals* took a markedly similar approach to discussing the play-development of animals, covering topics such as teasing, hunting, courtship, nursing and fighting, as well as introducing the concept of the divided consciousness in make believe.[48] He was describing the curious double state of game play; participating in make-believe yet responding as if it were real.

Other recommended books included Hermann Hesse's novel *Magister Ludi* (master of the game), also published under the title *The Glass Bead Game*. It is set in the future, in a fictional province peopled by intellectuals who dedicate their lives to learning and playing a complex and mysterious game.[49] Self also recommended John Cohen's *Chance, Skill and Luck*, which explored the impulses behind guessing and gambling in play.[50] Leslie Daikin's *Children's Games Throughout the Year* was lighter in tone, tracing the traditional games that children play in each season.[51] If we draw this diverse list together, the predominant themes and ideas presented to the students were play as human or animal development, play as power or competition, and play as analogical to life. This was coupled by the primary influence that was Caillois, with that compelling list that described reasons for play, instruments of play and outcomes of play. Taken with the fact that Groundcourse focused the students from the outset upon communication as a system, these themes created rich potential for exploring or manipulating group dynamics.

From these dark performances of play, it is evident that the students were developing a sense of the uneasy space between play and life, in which life was imitated but contorted, where the relationship between a playful 'sign' and its point of origin was not necessarily straightforward. As Mahai I. Spariosu has noted: '[...] in contemporary philosophy, play has been described as having an ambivalent ontological status, being both phenomenon and subjectivity or both behavior and intentionality.'[52] The sociological, anthropological and philosophical literature of play that emerged from the middle of the twentieth century all contributed to defining how play related to the human psyche, particularly in relation to the child. Our contemporary conception of play as a tool for development owes much to the immediate post-war years, particularly Gregory Bateson's essay 'A Theory of Play and Fantasy'.[53] Spariosu has recorded this influence, stating that the 'generation of 'child psychologists and educators' who came after Piaget believed it to have primary importance to cognitive development and function and:

> proclaimed it to be a decisive factor in the development of intelligence and the creative imagination. They support this contention by allying themselves with the recent research on play, conducted in communication, cybernetics, ethology, primatology, and anthropology. A good deal of this research has been inspired by Gregory Bateson's frame or metacommunication theory of play.[54]

Bateson's particular strain of anthropology brought together cybernetics and systems with anthropology and psychology. Bateson draws on observations of monkeys at play, noting that 'Paradox is doubly present in the signals which are exchanges within the context of play, fantasy, threat, etc. Not only does the playful nip not denote what would be denoted but the bite for which it stands, but in addition, the bite itself is fictional.'[55] Play then, is duplicitous, connected to the world but also a world of sign and symbol in itself. Bateson's used language that showed his interest in cybernetics, describing the information travelling between individual players as 'signals'. He wrote:

> If we speculate about the evolution of communication, it is evident that a very important stage in this evolution occurs when the organism gradually ceases to respond 'quite automatically' to the mood-signals of another and becomes able to recognize the sign as a signal: that is, to recognize that the other individual's and its own signals are only signals, which can be trusted, distrusted, falsified, denied, amplified, corrected, and so forth.[56]

In this passage, Bateson brought together the organism with the concept of communication signals, the latter derived from the new field of information theory.[57] This cybernetic apparatus was what made Bateson's essay a seminal contribution to play theory – he argued that play was in essence a flow of information signals, which meant in turn that it was unstable because of the subjective responses of players. Groundcourse, intended to be a cybernetic entity, often touched upon the same idea of information signals. In addition, by the second incarnation of Groundcourse at Ipswich, the subjectivity of play was becoming increasingly evident through experiment.

Bateson also argued that play was part of a field that consisted of '[...] threat, play and histrionics', which he suggested were often perceived as 'three independent phenomena contributing to the evolution of the discrimination between map and territory'.[58] He argued that they were not separate but enmeshed, using the example of childhood games that so often incorporated threat, fear and high responses into their fabric. He described how '[...] childhood behaviour shows that such combinations as histrionic play, bluff, playful threat, teasing play in response to threat, histrionic threat, and so on form together a single total complex of phenomena'.[59]

Bateson considered the understanding of 'map and territory' as an outcome of play, the ability to conceptualize the world into coded form and to understand both the connections and the differences between the two. It is certainly interesting that Bateson chose this analogy, given his explicit interest in cybernetics. Ascott's own interest in horizontal mapping developed at a time when the map was a recurring symbol in the arts, in philosophy and in the social sciences. It is unsurprising that some of the Groundcourse games unfolded upon the horizontal place, opening up mapped spaces or territories. In a sequence of photographs featuring Brian Eno and other unidentified students (Figures 6.11 and 6.13) a white table surface with a shallow rim contains a landscape of uneven white peaks. The raked mounds appear to consist of heaped flour or sugar; the players lean over the surface and throw a red ball at this white surface. The hip-level table and its strange topography spatially resemble the Fighter Control map, and the figures certainly bring to mind officers surveying the map below. Students given this task created boards that resembled maps or vector grids, formed topographies from piled flour and created enclosed and shadowy spaces of play. The position of game players – standing, leaning, looking down or over – indicated a level of active observation with the game boards forming a conceptual territory below them.

In a photograph from the operations room during the Battle of Britain, the Fighter Control officers watch the table surface with a sweeper stick held aloft, waiting for instructions to move a flight marker into place (Figures 6.6 and 6.7). In a relationship that opens up the treachery of play and what Bateson described as 'signals' between players, we see how closely the serious and tense business of military surveillance resembled play, while the Groundcourse games in turn appear meaningful, tense and serious.

Light and set

It is particularly notable that many of the student groups were interested in the creation of 'environments', using built structures, lighting and other props to build theatrical spaces. Ted Powell and his group at Ealing

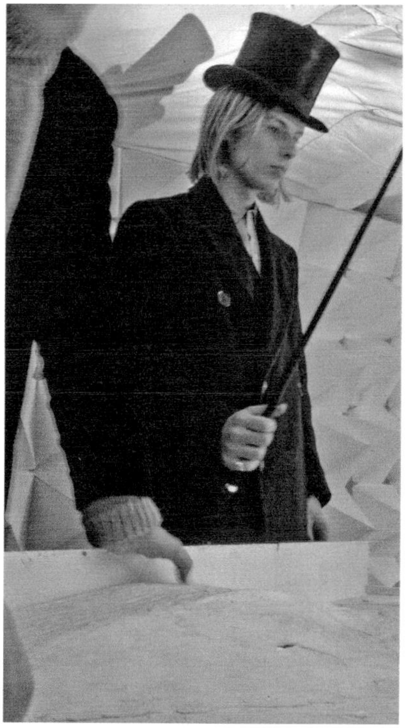

Figure 6.13 Roy Beston (1965) *Brian Eno: Groundcourse Behavioural Project.* Photograph courtesy of Roy Ascott

commandeered the whole studio for a week – they 'took over', building an installation of old newspaper tubes with a mesh ceiling and coloured lights in the corners. It had a false, sloping floor. For that week, people negotiated the studio on their hands and knees, crawling through tunnels.[60] It should be noted that these early experiments in installation were innovatory, taking place before either installation or immersive performance were established elements of the contemporary art scene. In a sense then, the post-Coldstream years in British art schools constituted a shift from art education that was firmly rooted in learning established techniques to an art education that was rooted in the principle of visual research and experimentation.

There are several reasons that the Groundcourse students found their way to installation or environmental modification. The first is the simple shift in focus that the course provoked, from object to system. The terms 'organism' and 'environment' were entrenched into the course, and the kinds of objects the students produced were always adaptable elements of an environment such as machines, diagrams, calibrators or props. In this systematized environment, it can be little surprise that many groups produced 'total environments' which their group or other participants had to negotiate. Townshend recalled that he left Groundcourse with a sustained interest in installation. He reminisced that he '[…] experimented with colour and semiotics, and a group of us built a large structure in our classroom, in which we intended to create an experience shed. My first attempt at installation sculpture, it felt like a fairground ghost train.'[61] In this tangle of recollections, it is patently clear that the students at Ealing had moved quite naturally into an expanded view of the visual arts, mirroring the course philosophy of art as an open-ended experience. Townshend's reference to a ghost train is a useful means to understand that the students were beginning to move into a more theatrical realm, creating performative experiences that resisted becoming spectacles through interactivity.

At this stage, Townshend thought his future would be in art and design, not in music, recollecting that the band '[…] felt like a side-project to me'.[62] When Ealing failed to gain the right to offer the new DipAD, he considered a move to another college. He remembers he was full of visual ideas:

> I was particularly interested in kinetic sculpture: installations, combining colour, lighting, TV screens and complex, coded music. All this, I imagined, would be interactive, brought to life by the computers Roy Ascott talked about.[63]

It is worth noting that Townshend remembers that even as a student he understood that the interactive art created during Groundcourse was implicitly connected to the advance of computer technologies. Further, he and his peers understood that with time computers were going to open out new realms of possibility for the visual arts.

In light of this, these proto-installations can be read as a conceptualized response to computers and the unfolding spaces of communication that Ascott believed them to represent by as early as 1961. This was the curious balance of Groundcourse – it frequently introduced students to sophisticated technologies and associated theories, many of which dealt with an increasingly technological future that cyberneticists in particular saw as inevitable. At the same time, the students used limited and often traditional materials to create environments – paper, card, found objects, string, household items, paint, wood, wire. In a sense, the sophisticated responses to communication technologies created by Groundcourse students and staff have been undermined by the fact the students did not directly employ technology in their work. Considering that these installations were produced collaboratively, it is also evident that the groups were cultivating a very different model of visual communication to the traditional maker-object-viewer communicative structure that had been the bedrock of art training in modernity. Groundcourse students exhibited a different understanding of the temporality of the work of art, treating it as an open-ended system rather than the endpoint of a process of production. This step into the present was the product of layered ideas, but it was certainly cybernetic theory that caused Ascott, his staff and his students to reposition their art in time, to abandon the static – or complete – in favour of art in process, as liquid and responsive as the flow of information between machines. Real-time art then, expanded the perimeters of the viewing experience. When Townshend's group wanted to build an 'experience shed', it was based upon the premise that the whole experience and all the elements from which it was composed would be the work of art.

The students had created a bunker of sorts – certainly, an enclosed, artificially lit space. The visible walls in Figures 6.11 and 6.13 are constructed from foil and a geometric grid of paper pyramids and the ceiling is covered with swags of white fabric, creating a total environment for their game/performance. The white and silver colour scheme has a B-Movie science fiction edge to it, but the most intriguing element is certainly the sense of enclosure the students have created. In the background, the colour saturating the wall constructed of white card pyramids is evidence of another Groundcourse class, in light effects. Brian Eno recalls taking part in a light-handling class in which students had to control the environment with lights, coloured filters, lenses and screens. He recalls that:

> The pyramid field at the back of the picture is made from folded card. The lights were not turned on in this picture, but what really happened was that there were coloured lights – red, green and blue – around the sides of the pyramid field and these produced very elaborate patterns of colours where they mixed. You can get a little bit of the effect from the fall of the white light on the surfaces, but this is of course much more complex, as there are different mixtures of colours.[64]

The coloured light that is captured in the photograph contributed to the environment of play. Light was from the outset one of the principal materials of cybernetic artists; as the decade progressed, the collaborative outputs at MIT's Center for Advanced Visual Studies would include responsive light installations by György Kepes, Otto Piene and Wen-Ying Tsai amongst others.[65] The New York EAT production of *Nine Evenings* had incorporated responsive light effects too, particularly in relation to sound.[66] We can see that this interest reproduced itself in the schools of art in the UK and across the world, as concepts of environmental design and control infiltrated the visual arts. In the context of cybernetic modes of art production, this can partly be ascribed to the fact that lighting circuits could be wired to sensors with relative ease by the 1960s, meaning changing light conditions could be used to create interactive spaces.

However, as Andrew Pickering has described, in the field of cybernetics, light had also been the subject of increasingly interesting research into psychiatric applications of light effects, mind control and trance-induction.[67] We can trace this from William Grey Walter's early light-responsive tortoise 'robots', which responded to the reflected shine of lights on their foreheads, to his subsequent discovery that flickering lights could induce fits or seizures. He had started work with the then little-known effects of strobe lighting in the post-war years – electric stroboscopes had become available during the war. He recollected that:

> Staring at the machine through closed eyelids [...] I remember vividly the peculiar sensation of light-headedness I felt at flash rates between 6 and 20 per second and I thought at once 'Is this how one feels in a petit mal attack? – of course, this could be how one can *induce* a petit mal attack.[68]

He discovered that it was indeed possible to produce attacks similar to an epileptic fit in non-epileptics. It was on these grounds that Harold 'Shippy' Shipton then built a flicker feedback machine, in which the light pulses were fired by the brain rhythms themselves. Andrew Pickering has offered an invaluable account of the ways in which cybernetic research was adapted – and incorporated – into both psychiatry and the psychedelic culture of the 1960s. From fantastical light shows at rock gigs, to 'dream machines' creating trance-like states in the home using stroboscopic flicker, by the end of the decade light was a tool for heightening and exploring mental states, often assisted by LSD.[69] However, much of the experimental work in light installation that took place in the UK and the US was well under-way before this psychedelic revolution. We can see its place in Minimalism, as Dan Flavin's early light work evolved from a distinctively anti-cosmic position. 'It is what it is and it ain't nothing else', he claimed, eschewing the more spiritual readings of his neon light installations.[70] Despite this claim, Flavin's work – read in the context of the structured object/subject interrogations taking place in Minimalism – intentionally manipulated the installation environment, intervened in it using light as substance. In 1967, when Willoughby Sharp observed that 'the art of light and movement is the only totally new art of our time', the interactions of the two had already produced some of the most compelling and technologically sophisticated work of the decade.[71]

In dealing with early light workshops in schools of art then, we need to consider not only the cybernetic use of light in terms of psychiatry and altered states, but also the fact that light is the perfect way in which to introduce students to the creation of controlled environments. It is transient, in flux – it is process. Ascott himself had been interested in interactive light for some years. As a first-year student at King's College, he produced a drawing for '[...] an electric light theatre with all the wiring', which is conceptually very similar to Pask's *Musicolour*, which had, unbeknown to Ascott at this time, been produced two years previously.[72] By 1971, when Ascott was working in Ontario, he had fully integrated lighting into the curriculum, planning a fully equipped light studio to be designed and led by Erwin M. Feyer, whom he had invited to join the faculty at Ontario College of Art.[73] He exchanged letters with Feyer, who provided detailed plans and also mentioned having '[…] had a phone conversation with György Kepes from MIT concerning the possibility of getting recommendations on someone who could offer a course on Structure' and that 'Kepes would also be available for a talk'.[74]

Ascott recalled a light class led by Feyer that began with a solitary candle in a dark room, working its way progressively through the atmospheric possibilities of light. Ascott described how they started '[…] with a darkened room and a candle and from there we go on, we obscure the candle, we have three candles, spotlight […] he developed a whole language of light. Light was his project.'[75] Like Ascott, Feyer was asked to leave Toronto. In Ascott's words: 'He was also decapitated when I was sadly […] when the mafia moved back in […] it was a really, really interesting workshop, coloured filters, everything you could dream of.'[76] From the early 1960s to the mid-1970s, light became a material approach and it was taught as such at the more experimental schools of art. For example, at Hornsey College of Art in the years 1965–8, there was a project called the Light-Sound Workshop. It involved John Bowstead, Dennis Crompton, Peter Cook, Roger Jeffs, Tony Rickaby, Martin Salisbury, Dante Leonelli and Ron Sutherland. The group worked, often collaboratively, on a number of projects and exhibitions involving light projections, film, animation, tape-slide programmes and sound. These included Miss Misty and the Tri-Cool Data, Aston University, Birmingham (1965), Ultra-Stellar Scanner, Brighton Festival (1967), Light/Sound Workshop, Oxford Museum of Modern Art (1968) and Time for a Change, Young Contemporaries, ICA, London (1968). Dante Leonelli was perhaps one of the less commercially motivated members of the department: he founded a 'Department X' at Hornsey which focused on light and sound. Leonelli saw light as a material to be manipulated, renouncing traditional artistic skills in favour of this.

The use of light is eminently theatrical. Eno's class experimented with slides of coloured glass – a method that had been used to produce special effects with gaslights during theatre production since the nineteenth century.

In the controlled, theatrical lighting of the Groundcourse games, we can see a profound connection to contemporaneous developments in cybernetics, which might help us read early practices in installation within a broader context of mental – or psychiatric – control. These games, both dark and flooded with artificial light, imply hidden, secret – even subterranean – space. There is an implicit parallel here between the enclosed environment of game play and the enclosed environment of the military bunker. While the bunker was an enclosed space, it represented extended geography; an ocean to be guarded from attack, a wide, open and dark space as extensive as the bunker was limited. The collision of the enclosed architectural environment with the extended space of surveillance came about through the displays, technological and analogue, which invoked this larger geography. Just as a game indicates imagined or parallel worlds of possibility, the ops room map projects a geography of open possibility in direct conflict with the physical confines of the space.

Democratic play

While British art schools have always been sites of conflict between tradition and progress, it was certainly more marked in the 1960s. Many of the new foundation courses were created by young, recently trained artists, often in conflict to the existing teaching agendas within the schools that hired them. In this way, foundation courses became the performance of values. We might, however, take this notion of the watcher further and apply it to the changing dynamics of the artist-viewer relationship in the 1960s. From formal Minimalist interrogations of spaces of display, to performance, installation and happenings, the work of art became defined as a site of observation. As we have seen, behaviourism – particularly the experiments of B.F. Skinner – formed a vital constituent of early cybernetics, and also formed part of Ascott's own field of interest in establishing Groundcourse. Behaviourism created and subverted performance. Its application within British art pedagogy not only generated new performative modes of practice such as game play, but it also presented a new philosophy of practice to students. There was no end point for them; there was no final object. Instead, objects, machines, diagrams and games formed part of a merging process of making and dissemination. However, despite this non-hierarchical approach to assessment, the dynamics of power certainly rose to the surface in the curriculum, particularly in the recurring themes of surveillance and behaviour modification. It is worth noting that by the end of the 1960s, worldwide student protest would stimulate further revolution in education as students called for more democratic models. Claire Bishop has argued that participatory art practices have created a situation in which '[...] vocabularies of social organisation and models of democracy have come to assume a new relevance for the analysis of contemporary art'.[77] Taking into account that trends of behaviourism and control in art education in the 1960s contributed to the development of participatory practices, we might consider that 'vocabularies of social organisation' are not only modes of analysis but of practice too.

The performance of the Groundcourse games discussed in these pages shines a light on another facet of the new cybernetic age – that the interconnected, architectural environments of modern warfare and the materials used to create them provoked an entirely new physical experience for those who occupied them. The ops room was a hub. It received data, which streamed in from the pulses of the radar and it was passed on via telephone, physically recreated on a map, cross-referenced and shared. Information appeared on screens. This in itself was the remarkable beginning of a phenomenon so familiar to us today as to be almost imperceptible: the change in spatial dynamics wrought by real-time information streamed onto screens and through the analogue displays. In the closed set of the art school studio then, an open network of communication became visible to those that participated in the Groundcourse experiments. Information flowed, transformed and took on new forms, through objects, performance and play. Communication was data, disembodied and changeable.

Notes

1 Winston Churchill (1949) "The Wizard War". *The Second World War Volume II*. New York: Haughton Mifflin Company, 337.
2 Willoughby Sharp 'Luminism and Kineticism' in: Gregory Battock (1995) [1968] *Minimal Art – A Critical Anthology*. Berkeley: University of California Press, 344.
3 Mark J.P Wolf (ed.) (2008) *The Video Game Explosion: A History From Pong to PlayStation and Beyond*. Westport, CT: Greenwood Press, xvii, 32.
4 Reproduced and discussed in: Helen Stuckey, Melanie Swalwell & Angela Ndalianis 'The Popular Memory Archive: Collecting and Exhibiting Player Culture from the 1980s' in Arthur Tatnall, Tilly Blyth & Roger Johnson (eds) *Making the History of Computing Relevant*. Selected revised papers from the IFIP WG 9.7 International Conference, London, June 2013. London: Springer 215–25, 216.
5 *Ibid*., 216.
6 Zach Whalen and Laurie N. Taylor (eds) (2008) *Playing the Past: History and Nostalgia in Video Games*. Nashville: Vanderbilt University Press, 87.
7 Robert Slater (1989) *Portraits in Silicone*. Massachusetts: MIT Press, 37–8.
8 For a sense of the reassuring tone of the Festival of Britain literature, see Jacob Bronowski (1951) *Exhibition of Science* (Catalogue-guide) Museum of South Kensington, 14.
9 For example, Caillois was quoted in curriculum documents at Ipswich. See: Roger Caillois (1961) *Man, Play and Games*. Illinois: University of Illinois Press.
10 Norbert Wiener (1961) *Cybernetics: Or Control and Communication in the Animal and the Machine*. Massachusetts: MIT Press, 162.
11 *Ibid*., 23–5.
12 Stuckey, Swalwell & Ndalianis (2013) 216.
13 Marcel Duchamp (1997) *The Complete Works of Marcel Duchamp Vol. 1*. New York: Delano Greenidge Editions, 72.
14 Jon Bird & Ezequiel Di Paolo (2008) 'Gordon Pask and His Maverick Machines'. In: Phil Husbands, Owen Holland & Michael Wheeler *The Mechanical Mind in History*. Massachusetts: MIT Press, 190.
15 *Ibid*., 192–3.
16 *Ibid*., 190.
17 Roy Ascott (1964) 128.
18 *Ibid*., 128.
19 Tim Stott (2015) *Play and Participation in Contemporary Art Practices*. London: Routledge, 2.
20 Claire Bishop (2012) *Artificial Hells: Participatory Art and the Politics of Spectatorship*. London: Verso, 8.
21 For example, Ealing and Ipswich have only scant records remaining of Ascott's short tenure. The existing archival material was mainly accumulated by Ascott, other teaching staff and former students.
22 For an account of this period, see D.W. Piper (ed.) (1973) *Readings in Art Education 1: After Hornsey*. & *Readings in Art and Design Education 2: After Coldstream*. London: Davis-Poynter.
23 The 'laboratory' model also appeared in non-pedagogical art projects in the late1960s. See Christopher Olsen (2000) 'The Art Lab Phenomenon in Great Britain 1968–1971'. Unpublished PhD Dissertation. University of Maryland.
24 Ascott (1964) 131.
25 Thompson (1917).
26 See 'Calibrator' chapter.

27 Ludwig von Bertalanffy (1969) *General System Theory: Foundation, Development, Applications*. New York: George Braziller, 32.
28 von Bertalanffy (1969) 4.
29 *Ibid*., 187–8.
30 *Ibid*., 49.
31 See *Metaform* chapter 28–9 for further discussion of Hamilton's use of grids in his exhibition projects.
32 See 'Field' chapter.
33 Kate Sloan *In conversation with Roy Ascott* (2012).
34 Kate Sloan *In conversation with Roy Ascott* (2015).
35 Lawrence Self. *Groundcourse Game Design Project*. Curriculum Document. Ipswich, 1965.
36 Roger Caillois (1961) *Man, Play and Games*. Illinois: University of Illinois Press.
37 Elizabeth Wright (2016) 'From the Life to the Locked Room: Applying the Chaîne Opératoire to the Pedagogy of the Copy'. *Journal of Visual Art Practice*, Vol. 15. No. 2–3, 245–60, 250.
38 Hester Westley (2007) 'The Year of the Locked Room' *Tate Etc*., Issue 9 (Spring).
39 See 'Control' chapter for a discussion of these courses in relation to British pedagogy.
40 Bateson (1955) 188.
41 Noel G. Charleton (2008) *Understanding Gregory Bateson: Mind, Beauty and the Sacred Earth*. New York: SUNY Press, 97.
42 Shanken (2003) 38–9.
43 Roy Ascott in conversation with the author. November 2015.
44 Lawrence Self (1965) *Groundcourse Game Design Project*. Curriculum Document. Ipswich.
45 Karl Groos (1898) *The Play of Animals* and (1901) *The Play of Man*. New York: D. Appleton and Co.
46 *Ibid*.
47 *Ibid*., 339.
48 Karl Groos (1898) *The Play of Animals*. New York: D. Appleton and Co.
49 Hermann Hesse (1943) *Das Glasperlenspiel*. See English translation: *The Glass Bead Game*. London: Owl Books.
50 John Cohen (1960) *Chance, Skill and Luck: The Psychology of Guessing and Gambling*. London: Penguin Books.
51 Lesley Daikin (1949) *Children's Games Throughout the Year*. London: Batsford.
52 Mihai I. Spariosu (1989) *Dionysus Reborn: Play and the Aesthetic Dimension in Modern Philosophical and Scientific Discourse*. London: Cornell University Press, 2.
53 Gregory Bateson (1955) 'A Theory of Play and Fantasy'. In: Gregory Bateson (1987) [1972] *Steps to an Ecology of Mind*. New Jersey: Jason Aronson Inc. 183–98.
54 Spariosu (1989) 196.
55 *Ibid*., 188.
56 *Ibid*., 184.
57 See Analogue chapter for a full discussion of information theory and also: Claude E. Shannon, (1948) 'A Mathematical Theory of Communication', *Bell System Technical Journal*. Vol. 27. No. 3, 379–423.
58 Bateson (1955) 187.
59 *Ibid*., 187.
60 Ted Powell (2017) Telephone exchange with the author.
61 Pete Townshend (2012) *Who I Am*. London: Harper Collins. 50.
62 *Ibid*., 55–6.

63 *Ibid.*, 55–6.
64 Brian Eno quoted in: Christopher Scoates (2013) *Brian Eno: Visual Music*. London: Chronicle Books. 24–5.
65 For an overview of the history of CAVS, see Melissa Ragain (2012) 'From Organization to Network: MIT's Center for Advanced Visual Studies'. *X-TRA*, Vol. 14. No. 3 Spring.
66 See Deborah Garwood (2007) 'The Future of an Idea: 9 Evenings Forty Years Later'. *A Journal of Performance and Art*, Vol. 29. No. 1 (January), 36–48.
67 Andrew Pickering (2008) *The Cybernetic Brain: Sketches of Another Future*. Chicago: University of Chicago Press.
68 Grey Scott Walter (1966) quoted in Pickering (2008) 76.
69 *Ibid*. Pickering (2008).
70 Michael Gibson (1987) 'The Strange Case of the Fluorescent Tube'. *Art International*, 1 [Autumn 1987], 105.
71 Willoughby Sharp (1967) 'Luminism and Kineticism'. In: Gregory Battock (ed.) (1968) *Minimal Art: A Critical Anthology*. Berkeley: University of California Press 317–59, 317.
72 Roy Ascott interviewed by the author. November 2015.
73 Letter to Ascott from Erwin M. Feyer detailing plans for the light studio, including detailed diagrams of ceiling grids and tracks, screening area and worktables. (1971) Archive of Roy Ascott.
74 *Ibid*.
75 Kate Sloan *Interview with Roy Ascott* (November 2015).
76 *Ibid*.
77 Bishop (2012) 7.

7 Synthesizer
Feedback in multimedia

Cybernetic psychedelia

In the late 1960s, multimedia events proliferated in British counterculture and particularly in relation to the booming music scene. These events often involved the new art of the light show, as well as other elements such as theatre, costume and burlesque. A trend towards creating environments – even architectures – of performance and interaction emerged, as psychedelia flourished. The fact that psychedelia was interactive, interdisciplinary and often cybernetic is little-known. In fact, psychedelic culture had a deep and sustained ideology of mind-altering experience that went far beyond its indelible association with hallucinogenic drugs. This fertile period at the beginnings of multimedia also saw a boom in art school rock, in art–music collaboration and in the use of technologies to enhance or distort performances. In the ultimate creative feedback loop, creativity flowed between art forms, enhanced by the possibilities presented by new technologies, including the synthesizer. When psychedelic counterculture was at its zenith, many individuals contributed to events for free. However, events such as those at the infamous UFO club and Roundhouse in London became a useful publicity tool for bands, given the number of journalists that were drawn by the sensation and scandal that surrounded the scene. Many bands of the period including Soft Machine, Pink Floyd, The Who and The Doors achieved fame and notoriety not only through their sound, but through their multimedia and theatrical performances. Other art forms – particularly lightshows – were vital to this, but the sense of participatory and collaborative practice behind countercultural happenings has been overshadowed by the popularity and fame of the bands who performed.

This final chapter pursues synthesis rather than conclusion, in that it outlines the dispersal of ideas from Groundcourse and other art schools into counterculture, as well as considering more broadly the rich and vital relationship between experimental art school pedagogy and multimedia in the music scene of the late 1960s. It will follow a few former students of Groundcourse, tracing their post-education experiences, projects and collaborations. In doing so, the chapter will draw out the enduring relevance of Ascott's cybernetic ideology and its long-term influence upon his students, as well as recovering some of the ephemeral and experimental collaborations that took place in Britain in the late 1960s. In keeping with Ascott's own philosophy of art, this is a decentred conclusion, extending outwards from the core ideas of cybernetics, behaviour and technology that have shaped this book.

The intuitive interdisciplinarity of artists and musicians in the 1960s generated art of the moment; it existed only through experience and then it was taken apart. As unpredictable and fluid patterns of light and sound danced across the walls of nightclubs, tents and warehouses, the dispersed cybernetic idea of communication systems merged with the countercultural pursuit of heightened consciousness. Despite its reputation to the contrary, the psychedelic scene was not about escapism; it was about experience and behaviour. It was a network of interrelated people and ideas, often simultaneously employing a network of interrelated machines, instruments and objects. Following on from one of the most potent periods in history for the British schools of art, a generation of graduates dispersed and synthesized.

The role of British art schools in the growth of the music scene in the 1960s is well-documented, given the number of art school graduates to migrate into music, including The Beatles, The Who, Roxy Music, Freddy Mercury, Ronnie Wood and Pink Floyd. In *Art into Pop*, Simon Frith and Howard Home's essential account of this phenomenon, the authors comment: 'Art school graduates are petit-bourgeois professionals who, as pop musicians, apply 'high art' skills and identities to a mass cultural form.'[1] Certainly, the rush of creativity in music technology and performance was defined by a more experimental approach to media and the same pursuit of audience interaction and response that was ever-growing in the schools of art. Frith and Home also observed that:

> The importance of the psychedelic movement was that in bringing the worlds of art and pop together it focused a question that had only vaguely worried the original art school musicians; what did it mean to be an 'artist' in the mass medium?[2]

This question crystallized, gradually, from experimental and intuitive work and play. The 'mass medium' was not in itself a known concept at the time – similarly, the term multimedia had yet to become common parlance. Instead, discussions amongst artists, musicians, critics, theorists and those in it for a good time hinged upon the experiences they were creating in the *moment*, and the creation of alternative spaces where they could achieve this. These alternative spaces and the multimedia happenings that took place in them represented the collaborative, alternative and essentially cybernetic spirit of the age.

Discotheque Interplay: a portable happening in the South of France

In the summer of 1967, the British artist Keith Albarn launched an extraordinary project that hinged around his *Discotheque Interplay*, a portable flat-pack performance venue that he had designed and created (Plate XII). There are numerous partial accounts of this event that have all been constructed from oral histories, press and biography extracts, mostly reproduced within the specific context of musical history. This is because Albarn invited the British jazz and rock band Soft Machine to perform as part of the event, and their subsequent fame has obscured the fact that the project was instigated by Albarn himself. This has been the fate of much of the psychedelic art that was woven into one of the most exciting, rebellious and turbulent music scenes of the twentieth century. It is a question of relative fame – despite the innate interdisciplinarity of the times, the phenomenon of music superstardom did not transfer to the other art forms involved in the scene.

Albarn had first studied architecture at Nottingham School of Art, before moving to London and studying sculpture at Hammersmith College of Art. He was refused a place at the Royal College of Art after getting angry during the interview process, something he puts down to the arrogance of youth. However, the rejection drove him to a series of decisions including marrying his then-girlfriend Jessica, who has been a lifelong collaborator.[3] Urgently seeking work and accommodation, Albarn was to discover that his unique double training in architecture and sculpture was of great value in the realms of theatre set and exhibition design. He was working and experimenting with materials that included various plastics, fibreglass, board and polystyrene, learning and experimenting along the way. This period of experimentation created interchange between the commercial work he was producing as a freelancer and his art practice. It can be little surprise then, that Albarn began to create architectures of experience: built environments that were intended to be both theatrical and multisensory, through which the spectator moved. Albarn used the money he earned in his freelance work to pay for his art projects, some of which also had funding from other sources. Albarn devised *Discotheque Interplay* after he was invited to contribute to a German beer festival taking place in the South of France, at the beach of Saint Aygulf, ten miles from St Tropez. It was Albarn that brought together a team of roughly twenty people, including Mark Boyle and Joan Hills and the members of the band Soft Machine.

As the few existing photographs show, *Discotheque Interplay* was an extraordinary piece of temporary architecture. At its centre was a fifty foot high tower, supporting rounded wings that together made a building that was 100 feet in length, with space for 500 people inside and a raised stage for the band. Its exterior comprised glossy block patterns in red, white and black. It lowered like a psychedelic spaceship on the golden sand. Albarn had overseen a chaotic build, in which *Discotheque Interplay*, transported by flatbed truck, had arrived before any of the installation team had bothered to lay the wooden platform for it. In order to raise the flat-pack building, the Basque assistants from the festival scoured the beach for muscle – the building was quite literally pulled upright by a large team of men in swimming trunks while all around them sunbathers watched with interest.

As noted, the members of Soft Machine and numerous other collaborators had also fetched up at the beach. Amongst their number was the former Groundcourse student John Bonehill.[4] The band had agreed to first take part in Jean-Jacques Lebel's production of *Desire Caught by the Tail*, the surreal, chaotic and provocative play written by Pablo Picasso in 1944. The producer Gorgio Gomelsky had arranged the pop-up performance at Saint Aygulf, and perhaps other gigs that were later aborted. Gomelsky later recalled that he had been approached by '[...] a French promoter, Jean-Pierre Rawson [...] asking if I knew a band willing to appear in a Jean-Jacques Lebel Happening-like production of Picasso's only play "*Le désir attrapé par la queue* [...]" I thought SM would be perfect for that.'[5] While the details of the trip are vague, it is evident that the performance of Picasso's play aroused suspicion and censure from the local mayor. Bonehill recalled that 'Picasso's only play, "Desire trapped by the tail" & "Happening" with the Soft Machine, were held under a blue circus tent in Cogolin, South France, because the Mayor of Saint Tropez threw us out of the Festival for being too radical.'[6] In London, Bonehill had worked on the famous light-shows at the nightclub UFO and had worked previously with Lebel.[7]

In other accounts, the band were shut down quickly by the mayor Jean Lescudier, who apparently said they made '[...] his coastline look like a 'pigsty'.[8] He also apparently believed the Picasso play was '[...] nothing more than a pretext' and '[...] the "happening" which will follow may stir up trouble amongst the population [...]'.[9] From the existing facts then, the band performed in and after Lebel's production of Picasso's play, before setting up the *Discotheque Interplay* on the beach, in close proximity to the beer festival taking place in Saint Aygulf. What transpired was a very visible, audible and unintentionally disruptive act of avant-gardism, in close proximity to the more traditional and tourist-friendly offerings of the festival itself. Within the building, the band played on a stage raised above the audience.[10] Keith Albarn described an acrimonious end to proceedings:

> We underestimated the impact on the locals and trippers. The discotheque, which held some 400 souls, was welcomed by the younger punters but was viewed as somewhat extreme by nearly everyone else and met with stiffening opposition, leading to instances of blatant sabotage – cutting our mains cable, etc. The Soft Machine played heroically – and could be heard and seen for miles along the beach! But overall it was a painful experience. We had to bribe our way on and there wasn't enough of the take left for us. In the end we were obliged to blow the whole thing up on site to avoid being charged export taxes. Our remaining kit was then impounded at Calais by customs![11]

Albarn also remembers the solemn violence with which *Discotheque Interplay* met its end – several officials arrived on the beach, dressed formally in black suits in order to witness this. They ordered the group to take the building apart, and then to break each of the panels into pieces.[12] Alarmingly, Bonehill recalled that '[...] the farmer blasted the generator, with a shotgun!'[13] In *Downbeat* magazine, Mike Zwerin commented that '[...] the wrong element (penniless) was hanging around the discotheque but not drinking beer'.[14] He described the visual power of the event: 'Enveloped in Mark Boyle's moving, abstract projections, floating in many colors on top of them, the Soft Machine presents quite a sight onstage, though not an unusual one in our time.'[15]

If we put aside the friction and the abrupt end to the tour, what is left is the perfect example of the dispersal and convergence of cybernetic ideologies from art school to rock and back again. To start with, Albarn's art school training had led him to create what we can read as alternative spaces for the countercultural movement. Sometimes these were self-referential, contained by the notion of an 'experience space' and framed within the context of a gallery or nightclub, but others such as this took place in more challenging locations. Soft Machine were named for William S. Burroughs' novel of the same title, which Elda Danese describes as representing '[...] in a frame without hope, his obsession of control that post-technological society wields on humans and the social body'.[16] The band worked closely with light artists, to the extent that Mark Boyle was a salaried member of the band during the late 1960s and his *Sensual Laboratory* was an integral part of their performances.[17] Also, their crew that summer included Bonehill, who had been interested in light as a medium since his time as a student at Groundcourse. In his words: 'We experimented a lot with light manipulation. Di-achromatic spots had just been discovered, & much of this would show up later for me.'[18]

Mark Boyle's work was not a background. It was part of the fabric of Soft Machine performances in the period. In addition, we must take into account the extraordinary temporary architecture that Albarn had created in the form of the *Discotheque Interplay* itself. Boyle described it:

> That summer we went off with the Soft Machine to play in a translucent white plastic pavilion on the beach [...] We were to project from inside on to the whole skin of the pavilion, so that the light show would be seen from outside as well as inside. [...] It was part of a festival. The idea worked. At night it lay like a great space-ship on the beach, made out of rivers of colours and movement like erupting stained glass. With the shattering sound of the Soft Machine it was a fantastic success.[19]

Boyle's description of the translucent plastic 'skin' of the *Discotheque Interplay* architecture and the dancing patterns of his own projections upon them highlights the way in which the performance was both enclosed and exposed. The sound rippled along the beach and the low, strange building glowed and flickered with oily light and with the dancing shadows of enclosed figures. *Discotheque Interplay* consisted of layers of collaborative activity; the design and execution of the building itself, the planning and execution of the light show, the rehearsal and performance of the band. The work of art existed in combination and at the point of experience, as oil patterns formed and dispersed like mysterious organisms, against the insistent beat of the music. Mike Ratledge, the keyboard player for Soft Machine, wrote that:

> Mark Boyle's 'events' are content with a direct presentation of the reality that already exists, with no self-interposition from the artist [...] Whereas 'happening' implies agency, 'event' is the effect of something happening; to perform a 'happening' it is necessary to act, but one cannot act an 'event.' It is sufficient to realize the fact that a 'happening' has occurred. An 'event' is a discovery of what is happening, a 'happening' an active invention-fact as against act.[20]

Bonehill reminisced that: 'Mark Boyle was experimenting with "microscopic" projections such as pond water, & blood, saved in buckets from the chickens sacrificed on stage [...].'[21] Certainly, Boyle had been working with blood, semen and other substances during the period. Daevid Allen, the co-founder of Soft Machine, described the development of Boyle's biological light work:

> Mark Boyle had evolved his early light show at Mike Leonard's Sound/Light Workshop at Hornsey Art College, the same play that Joey Gannon's ideas were developed. Boyle operated the stage lights and sometimes played tricks on the bands, like making green bubbles emerge from their tightly- stretched flies – a trick he liked to play on Roger Waters – but he was only one of the many lightshow operators and the Floyd usually had their own spectacular show.[22]

Emphasizing the place of art schools in encouraging the development of lightshows and other light experiments in the visual arts, Allen's words give an insight into the playful opportunities that the medium of light opened up, performing in real time. He also provided an essential description of the visual experience of the light shows. He wrote: 'The blobs and bubbles of the lights were absorbed into each other or divided like the school diagrams of amoebas. One's whole vision was filled with pulsing cellular forms, like being inside a plant or your own body, the sap rushing, being borne along by the relentless rhythm.'[23] Allen's description brings to mind the biological abstraction that was still entrenched in British art school teaching. It also recalls Herbert Read's description of Herbert Read's assertion that art '[...] arises, like a green sap; like a seminal fluid, it issues from the body, and from the body in an unusual state of excitement.'[24] While Read would certainly not have anticipated the use of seminal fluid in a work of art, the collision of biological abstraction, projection and performance in these late 1960s projects certainly extended from these earlier preoccupations in British abstraction.

Bonehill also recalls the violence and sexuality of the performance of the Picasso play in high detail:

> During the 'happening' [the chickens'] throats were slit and their bodies swung around on the stage [...] Next came paint, and tomato sauce, under the projected images of bubbles and blood corpuscles, and Rita Renoir urinating on a very willing volunteer [...] The play was narrated, and the face of the narrator TV-projected onto a 4ft. Styrofoam head on the stage, and much more [...].[25]

Bonehill's words illuminate the curious place that lightshows held within happenings, suggestive and physical. It was a fluid medium, viscous liquids pressed between glass slides and once the heat of the projecting lamps warmed it, pulsing with life. In using blood or semen, Boyle and other lightshow artists of the era created a pure connection between material and vision; it was, in fact, biological material that danced with mesmerizing life across hung white sheets or plastic walls. It is also worth noting that Boyle himself worked in secrecy; Allen reminisced that at UFO, Boyle '[...] worked inside a tent so nobody could see what he was doing. Some said he used his own fresh sperm mixed with the colours and other liquids and fluids.'[26] In the open spaces of warehouses and basements that were the venues for countercultural happenings, filmy fabric subdivisions of space were often employed: tents, pulsing fabric balloons, sheets creating walls and spaces within spaces. Similarly, Albarn created architectural spaces within spaces; this was an architecture of experience and sensation. Allen saw it as a shroud of secrecy cultivated by Boyle as a 'liquid lightshow alchemist' whose '[...] combinations of liquids [...] sandwiched between the twin glass lenses, that began to alter as they were heated by the projector lamp, were his professional secret'.[27]

From these evocative memories, we can begin to set Albarn's St Tropez happening and its extended network of contributors in context. It was a period in which countercultural idealism, technology and interactivity culminated in integrated psychedelic experiences. The history of psychedelia has been somewhat constrained by its association with LSD and other drugs, when in fact they played only a small part in the agenda of mind-altering at play. As Simon Rycroft has so precisely described, the many artists and musicians who produced multimedia experiences had a broad and conceptually sophisticated agenda:

> Many of the performative innovations were directed against linear representational practices and attempted to create microcosmic, multi-sensory versions of the macrocosmos. These microcosmic 'environments' it was felt, would invoke the discovery of the 'true self'. The message in

> these activities was that the self was fundamentally networked, not only to other subjects in the world, but to the objects of nature and technology within and beyond the world.[28]

Ryland's conception of the microcosmic and the macrocosmic in happenings situates the 'non-linear' practices of psychedelic artists against the same issue of consciousness in a networked world that shaped Ascott's agenda in the early 1960s. That is to say, Groundcourse introduced a generation of British art students to a vision of an interactive, responsive, networked and intermedia mode of arts practice. There was, writes Ryland, '[...] departure from old ideas of matter and form and as we move into the post-war period an increased emphasis on the elemental and primitive'.[29] He noted that artists and theorists '[...] developed tactics and aesthetics that would capture, visualize or materialize these new and fashionable conceptions of nature and the cosmos – to make them known to human senses'.[30]

Groundcourse was the earliest example of a defined systems art curriculum in the UK, but the more experimental art schools across the UK also integrated light, sound and interaction into their curriculums to varying degrees. For example, David Mellor has noted that the Hornsey College of Art was at the forefront of this trend too:

> In the second half of the Sixties a new emphasis around an enfolding public art of light, of projections and spectacle, underpinned the appearances of multi-media events such as those produced by the Hornsey Light/Sound Workshop. In November 1967, at the Museum of Modern Art, Oxford, the Workshop programmed a series of light and music experiments in a space sixty by thirty feet by twenty feet; '[...] on a scale large enough to properly engulf the spectator.[31]

At the Hornsey workshops, a generation of art and design students were introduced to experimental and collaborative live practices. Given that these workshops hosted Pink Floyd's first performances, the Hornsey workshops certainly had an important part to play in the burgeoning alternative scene and its wired in, electric approach to creating mind-altering experiences. David Mellor has also recorded that '[...] Gillian Ayres had used filtered lights projected onto still-lives, during her year teaching Foundation students at Corsham Court during 1965–6 [...]', subverting a traditional training activity with artificial colour and light.[32] Taken together with Ascott's light workshops at Ealing, Ipswich and then in Canada and the United States, it is evident that light was an emerging medium in the post-war art school.

Materials of light: Boyle and Hills' *Bodily Fluids and Functions*

Mark Boyle and his future wife Joan Hills had, the year before St Tropez, presented a lightshow entitled *Bodily Fluids and Functions* (Figure 7.1),

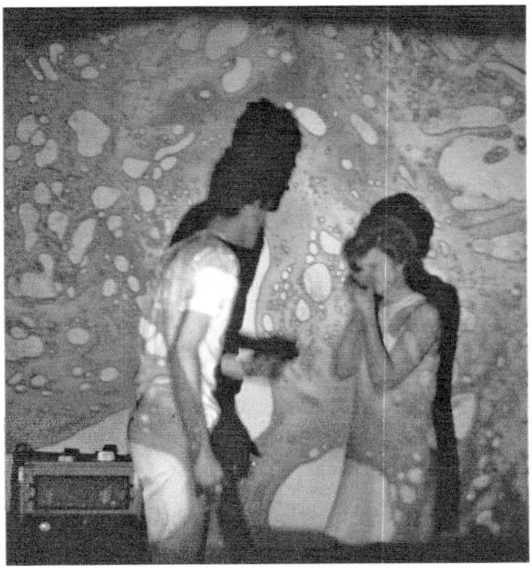

Figure 7.1 Mark Boyle & Joan Hills performing *Bodily Fluids and Functions* (1966). Photograph © Boyle Family

which had showcased their growing interest in experimenting with the visual qualities produced by liquids derived from the human body.

As described by Jasia Reichardt in the *New Scientist* in 1973, Boyle and Hills had extracted 'different bodily fluids' from 'one person and shown on a screen with a microprojector to the accompaniment of an analogous sound. The fluids included saliva, tears, semen, urine, sweat and blood.'[33] Zwerin describes these bodily fluids as having been '[...] projected onto a large screen together with electroencephalogram and electrocardiogram responses of a couple making love, while the sounds of the bodies were amplified throughout the auditorium'.[34] Boyle himself playfully commented: 'In the context of everything, anything is a fair sample, or, to put it another way, nothing is a fair sample.'[35]

Reichardt's words didn't quite capture the raw physicality of the event, something better extracted from Boyle and Hills' own notes on the project, worth reproducing in full:

FLUID ACTION SOUND REACTION PROJECTED ON TO A SCREEN

1 Catarrh Coughing up Contact mike on throat Examine with microscope dyes
2 Snot Blow nose etc. Contact mike on nose Examine with microscope dyes
3 Saliva Kiss to arouse saliva possibly eat Contact mike in mouths of kissers or on throat if eating Examine with microscope dyes separate starch etc.

4 Earwax Extract Melt and examine,
5 Tears Extract Tape children crying adults howling Crystalize tears in microprojector
6 Urine Piss in can Contact mike on can Sugar test on urine
7 Sweat Exercise furiously Breathing and heartbeat Examine with microscope dyes
8 Blood Extract with hypodermic Loud breathing and heartbeats Examine with microprojector
9 Sperm Extract by copulation or masturbation Attach electrocardiogram and electroencephalogram Televise oscilloscopes of equipment, project using T.V. projector Maybe silence Maybe climatic heart and breathing sounds Project live sperm in microprojector
10 Gastric Juices Swallow sponge on the end of a string Pull up Contact mike on stomach Examine with microprojector 11. Vomit Take emetic Contact mike on stomach and throat Examine with microprojector'[36]

This is an uncomfortably visceral list, focusing on what we might describe as the *behaviour* of the fluids produced and expelled by the human body. The heightened recorded sounds of cries, pants, groans, retching, trickles and snorts almost subverted the visuals, which floated and pulsed with the abstract beauty of organisms. In this sense, the event was bodily and physical in the extreme, visually not unlike the tours of the interior of the human body that have since been made possible by keyhole technologies. Recalling Boyle's light shows, Daevid Allen commented that 'People sometimes just stood open mouthed'.[37] After attending Boyle's screening a the Sandon Theatre in Liverpool and enjoying the gentle beauty of the imagery, a critic warned: 'Watch it, Mr Boyle, you are getting dangerously close to art; through tonight's X programme (Son et Lumiere for bodily fluids and functions) might restore the balance.'[38]

It is worth noting that early light shows were inherently experimental when it came to material – it was a case of working out how liquids would appear in projected form and what factors altered the visual qualities of liquid. The predominant factor was heat; by exposing inks or bodily fluids to a heat source before performance, or by using the heat of the projector lamps during the performance, liquids could be made to move, in viscous patterns, or to jump erratically. In the case of the spermatozoa that Boyle sandwiched between glass slides, it was the movement of life itself that he projected onto the wall. In the first showing of *Bodily Fluids and Functions*, Boyle and John Claxton adjusted the focus until suddenly, the image sprung to life: 'There were so many, all swimming in the same direction, millions and millions. A flash of eternity.'[39] He also recalled that in Bristol, everybody was surprised to see that tears crystallized on the screen.[40] Viewed from the twenty-first century, Boyle and Hills' project was essentially a systems treatment of the body as a series of expulsions, liquid, solid and aural. It played upon the intimacy of performance, while also playing upon the very contemporary issue of scale; the microcosmic world of the body enfolded the audience. In this sense, the popularity of the *Sensual Laboratory* in British counterculture is easily understood; it was certainly a mind-altering experience.

A multimedia blueprint: the launch of the *International Times*

Another crucial moment in the history of multimedia performance took place on 15 October 1966 at the launch party for the *International Times*, a magazine that was to become the voice of British counterculture. At 2 a.m., 2,500 people were still queuing hopefully outside a dilapidated circular warehouse in Chalk Farm, London. A further 1,500 people were inside the warehouse, which had been transformed into a space of shimmering illusion, with two lightshows, inflatable sculpture and balloons and painted, costumed partygoers. Sugar lumps were dispensed to guests near the door, who all assumed they contained hallucinogens. They were in fact only sugar, but a kind of placebo trip effect took hold for those who hadn't indulged in any other substances. 'Bring your own poison', the flyer and poster had declared, '& flowers & gas-filled balloons & submarine & candy & striped boxes & ladders & paint & flutes & feet & ladders & locomotives & madness & autumn & blowlamps & […] !?!?!?!?!?'. There are many well-worn anecdotes about this event, from the minimal length of Marianne Faithfull's nun's habit to the orgiastic tangle of couples making love in the shadows at the back of the hall.[41] The event featured layers of collaborative activity, particularly the interaction of the visual and the aural. Sculpture, including balloons, made up part of the 'stage setting' for the event, often treated as another surface which could be used for light displays.

The Boyle Family's *Sensual Laboratory* was one of two lightshows. Jack Bracelin, who operated light show company called Fiveacre Lights, was set up opposite. In his account of the career of Pink Floyd, Julian Palacios described the scene:

> Boyle wore plastic gloves and goggles, mixing mad chemicals in glass petri dishes over the blazing projector[…] Dermot Harvey (who Miles called 'an errant biochemist, discoverer or many immiscible liquids') projected three-dimensional liquid onto helium-filled balloons, which were forever drifting towards ceiling fans and lights.[42]

This description draws out the hands-on experimentation that took place at the Roundhouse and other countercultural venues, with scant regard for the kind of health and safety concerns that are imposed upon contemporary performance events. It is also worth considering that the artists, designers, musicians and engineers who created these events worked with quite rudimentary set-ups, including lots of trailing wires (Figure 7.2).

The setup of plugs and cables in Figure 7.2 is engulfed by shadow and overlaid by patches of bright, hectic light. In the top left of the image, a raised

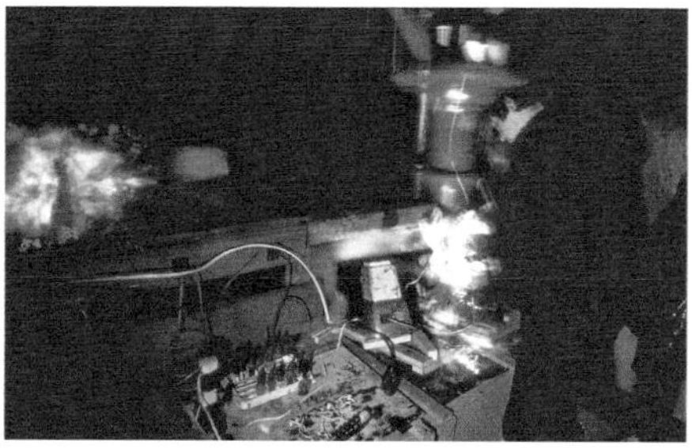

Figure 7.2 Roundhouse, Joe Gannon Light Projecting at the Launch of the *International Times*. Photograph © Adam Ritchie

fist is visible in a patch of glowing light overlaid with scribbled motion lines. We can assume that this hand was holding a lamp aloft, to create moving light effects on the walls, the audience and performers; this was a common tactic in early light shows. The light also moved over sculpture, balloons, amps, lighting rigs and musical instruments, all elements of a holistic environment of alternative participation.

One of the light artists involved in the event was Joel Brown. Brown had, a few days before, been introduced to Pink Floyd when they played their first gig at the Light/Sound Workshop at Hornsey College of Art. He had promptly been invited to bring some slides to the launch event in order to provide a light show for their performance.[43] Brown had recently moved to London from the United States, where he had '[…] a friend who made a fantastic light machine that projected a microscopic image that was incredible […] he went with crazy […]'.[44] In the UK, he experimented with inks, discovering that once heated they looked much better in projections. He had not seen the space before the opening night, when he stepped into a space that had in its centre 'A huge pearly iridescent balloon with projections on, like a hot air balloon-bubble […]'.[45] This balloon was inflated in the centre of the space, a filmy, pulsing skin, patterned with dancing light.

The stills from Brown's projections (Plate XIII) offer a glimpse into the realm of heady beauty that these light shows created for the willing audience, enhanced for many by alcohol or other 'poisons'. Brown describes a scene of wonder in which 'everyone stopped and watched the lightshow on them […] they sat on the floor and watched the show'.[46] The audience themselves (Figures 7.3 & 7.4) are tightly packed into the space, standing and watching the show with a still, almost uneasy concentration.

In Plate XIII, there are soft blobs and precise patterns like dendrite crystals and leaves. There are points of light like constellations and glowing embers. There are precise webs of craquelure over bubbles like lava. This extraordinary image records the embedded biological language of the light show, in which the patterns of nature were reproduced in shifting form and colour. It is compelling that these experiments, which often combined

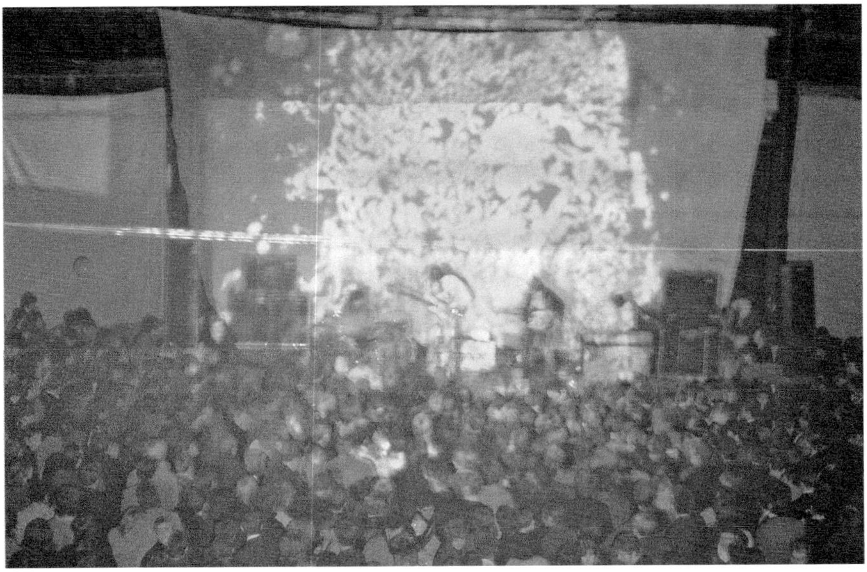

Figure 7.3 Pink Floyd Performance at the *International Times* Launch Party. Photograph © Adam Ritchie

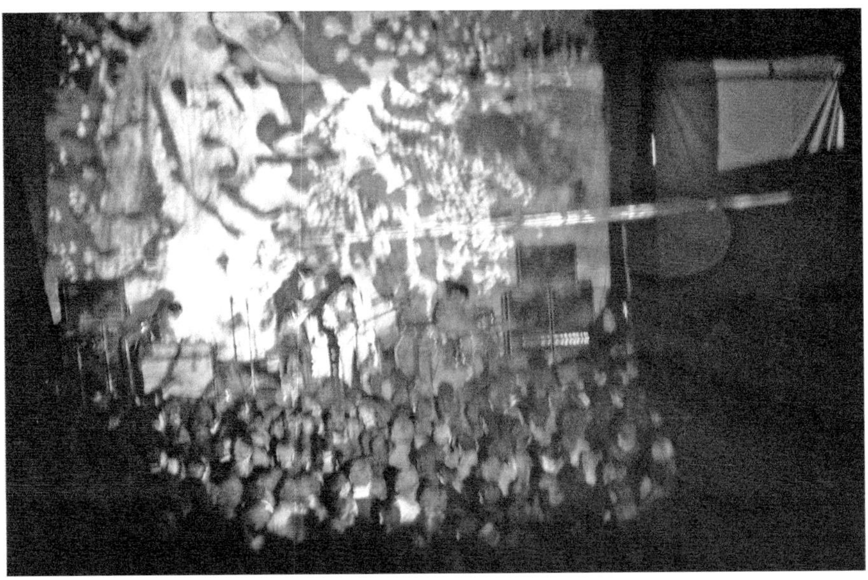

Figure 7.4 Joel Brown (1966) *Rebirth* created for Pink Floyd, performance at the launch of the *International Times*, Roundhouse club

chemical and organic matter, reproduced in the branching, cellular and crystalline patterns of life itself. It is unsurprising then, that light shows became so ingrained in 1960s happenings; the Alice in Wonderland effect of scaled-up microphotography in motion was like a trip into the microscopic world. In a review of the event published in *Melody Maker*, Nick Jones commented: 'The Pink Floyd have a promising sound, and some very groovy picture slides which attracted far more attention than the group, as they merge, blossom, burst, glow, divide and die.'[47]

Brown's projections were part of Pink Floyd's performance, and the light element certainly changed the performance dynamic at a time when new bands were often ignored by audiences, heckled to cover hits by other bands or treated as background. Brown remembers that at the Roundhouse gig:

> Some were standing still, some were bopping around a bit [...] it was very weird. Very strange indeed. Especially if your head was buzzing from imbibing things [...] It fit the mood and the time. The music and the light show complimented each other perfectly.[48]

Light played a crucial role in recasting the rock performance as progressive and creative, a visual and aural experience that might include unpredictable behaviours and shock.

Boyle has pointed out that the unity between light and sound was a trick of the mind: '[...] it was not synchronised with the music, though your brain forced the two together, making you think it was, which created something magical.'[49] Some light artists attempted some kind of synchronicity but really, the technologies were not advanced enough to allow for this. Instead, the audience created their own relationships between light and sound, seeking pattern out by nature. Allen commented that although Boyle's light shows '[...] could not be logically programmed, his lights synchronized with our stops, starts, peaks, and lows, as if it had all been pre-organized by a wizardly Atlantean reincarnate'.[50] The reference to Atlantis also demonstrates the broader reach of Michell's *View Over Atlantis* which had also interested Ascott.[51] For Allen then, the beguiling synchronicity of light and sound that Boyle's work created had a quasi-mythical power, which is why he describes him as a magician or an alchemist. Zwerin described Boyle's projections for Soft Machine as '[...] somewhat more modest' but opined that '[...] they add an exciting visual dimension to the music. Unlike most other light shows, he uses no stills and no objective images of any kind. The light and movement, formed by liquid chemicals, are determined by chance factors.'[52] Zwerin quotes from a 'scholarly paper' apparently written by Michael Ratledge about Boyle, in which he claimed that Boyle's '[...] presentations make it possible for the spectator to rediscover the 'esthetic' aspect of our environment that has become hidden by accretions of use and habit and to become aware of [...] environments that were previously inaccessible to us'.[53] This claim chimes with the idea of the rediscovery of 'hidden' ancient landscapes that were so prevalent in British counterculture.

Whether planned or not, light and music shows were an associative experience, patterns produced in the mind. However, some plans for synchronicity did succeed:

> [When Boyle] [...] put acid on perforated zinc, it would immediately attack zinc with ferocious action: pieces of zinc would fly off the screen with bubbles exploding when Mike Ratledge, keyboard player for Soft Machine, was going mad with the organ, or Robert Wyatt was doing his amazing drum solos.[54]

Bubbling zinc and acid created destructive, violent reactions that leapt and shuddered in time with the music. It is worth noting the similarity of these light projects to Metzger's practice too; as Mellor has noted, 'Metzger also continued his public demonstrations, in January 1966, by using liquid crystals heated, magnified and projected, first in the windows of Better Books on the Charing Cross Road, giving this 'demonstration' the cosmic title *Earth from Space*.'[55] Better Books was the site for many interdisciplinary art events in London during the 1960s.

It is worth recounting the other light show that took place at the launch of the *International Times* – Jack Bracelin of Five Acre Lights had also been running a projection. Bracelin often hosted happenings in London in the period, including working as a light producer for the famous UFO club. Rycroft has recounted that he also: '[...] took on the job of projecting Pink Floyd's lightshows as well as those of other psychedelic groups The Soft Machine and The Social Deviants.'[56] Bracelin was an interesting figure. He was High Priest of the Bricket Wood Coven, which was based at Five Acres, a naturist colony near Watford, since the 1940s. According to Barry Miles:

> Fiveacres was his psychedelic nudist colony near Watford, which consisted mostly of local teachers living in caravans. The wooden clubhouse had a 'trip machine' consisting of an electrically powered wheel on the ceiling from which strips of silver Mellonex hung down to the floor. As the wheel slowly turned, the assembled tripped nudists watched the flashing colours to the accompaniment of a very scratched copy of Zappa's *freak Out*. The Pink Floyd played there on November 5, 1965, on their way back to London from a concert at Wilton Hall, Bletchley.[57]

Five Acres still operates as a naturist site today, although one imagines that the trip machine might be a thing of the past. Rudimentary though it was, the regular pulsing light of the wheel was a sophisticated idea, connected to the growing interest in light pulses as a psychiatric tool. Bracelin himself had started his light shows with the intended audience of patient groups at mental hospitals. Miles records that '[...] the best of these was made by a girl patient at Knapsbury mental hospital' and they caused some strong reactions:

> The slides consisted of bright heaving masses of colour and produced amazing emotional reactions, tears and often a state of disturbance which lasted for days. Because of these reactions some of the hospitals he visited decided that his shows were 'too loaded' emotional and therefore stopped them. The effect at UFO was sometimes equally hypnotic and people danced for hours surrounded by the swirling shapes and points of light.[58]

From a contemporary standpoint it seems extraordinary that light shows such as this could be used within therapeutic settings in such a casual manner, but as noted, the use of light as a psychiatric tool was extremely prevalent in the 1960s. I have already described the overlap between psychiatry and cybernetics in Britain but in extension to this, the work of R.D. Laing and his contemporaries often hinged upon similar ideas to those that drove counterculture. As Pickering has noted: 'Laing's interest in LSD coupled with his psychiatric expertise gave him an important place in the London drugs scene of the sixties.'[59] Pickering also recounts that friends brought Syd James to Laing several times during the period in which he took a lot of acid.[60] As Laing himself noted, LSD was initially considered as a substance that provoked a temporary psychotic state:

> LSD-25 was originally regarded as a psychoticomimetic substance. I propose that this biochemically induced 6–12 hour trip has its natural analogue in what I suggest be called a *metanoiac* voyage (from metanoia: change of mind). The nature of the metanoiac voyage may be 'good' or 'bad', largely depending on the *set* and the *setting*.[61]

Laing's interest in creating alternative spaces for the treatment of mental illnesses is not dissimilar to the pursuit of alternative spaces – basements, beaches, caves, tents, architectures, caravans, warehouses – for countercultural happenings. In both contexts, the *set* had a defining role in experience, an interesting overlap with regards to the ever-more environmental concerns dominating the visual arts. Laing thought the *setting* had the power to create either a 'voyage of discovery into self of a potentially revolutionary nature' or 'a catastrophe: into a pathological process from which the person requires to be cured.'[62] Laing was also interested in group dynamics in therapeutic settings, commenting that: 'A change in one person changes the relation between that person and others, unless they resist change by institutionalizing themselves in a congealed professional posture [...] any transformation of one person invites accommodating transformations in others.'[63]

Bracelin's experiments in light shows for psychiatric hospitals and for night clubs and other happenings illustrates the fact that a preoccupation with altered consciousness traversed social, cultural and medical boundaries. Mellor described this interplay, suggesting that the invention of *Flicker Machines* had been based upon 'proto-Underground' experiments of Brion Gysin, Grey Walter and Ian Sommerville', and that they were '[...] the prototypes of the light devices being used in the mushrooming 'Underground' clubs.[64] Barry Miles noted that 'Throughout the Fifties and Sixties there were many experiments done to examine the effect of stroboscopic lights on early experimenters with LSD', which in turn demonstrated a definite impact upon their subjects. Miles also pointed out that '[...] the Pink Floyd were the only ones facing the light projectors, and Syd was the only one on drugs', so he was more susceptible to the trance-inducing effects that the light shows provoked.[65] Miles added that the cycle of flashes affected the potential impact, which may be why sometimes Syd was disoriented and other times showed 'flashes of brilliance'.[66]

The techniques that Bracelin employed in his light shows were very similar to Boyle's, in that he used a mix of organic and chemical viscous liquids sandwiched between glass slides, which he would squeeze with long-nosed pliers in time to the beat of the music. Some of the slides bore drawings in wax crayon, which resisted the mobile patterns of oil and added another layer of static imagery to projections. The countercultural light show migrated from the walls of underground clubs to the walls of mental health facilities and back again, using rudimentary and experimental combinations of materials to blend sound and light into trance-inducing patterns. As Rycroft has observed, it was not only hallucinogens that induced mind-altering experiences; more important were the 'electronic technologies' that were used to '[...] create "environments" in which a collective revelation could be facilitated.'[67] In an interesting measure of the quasi-therapeutic experiences at happenings, Mark Boyle reminisced about the following incident, after his:

> [...] brother was in a mental hospital in Glasgow, but he got himself down to London and came to stay with us. We took him to UFO, partly thinking it might help him in some way. There were hundreds of people there and he soon disappeared into the crowd, and we didn't see him for two days. We later learned that he met a girl who turned out to be a psychiatrist. He told us how he'd had a dream 'about a place with pictures going all the way around the room. Not pictures like you make, but pictures of moving colour, swirling and swimming, red, purple, green, orange. You would have loved it, it was kind of magical'.[68]

Pink Floyd and UFO

As the 1960s progressed, the band Pink Floyd worked with Boyle and other artists and designers to refine and experiment with the light show concept, seeking out interplay between sound and light. They had a team of light-operators that consisted of Peter Wynne-Willson, Susie Gawler-Wright and Joey Gannon. Miles has recorded that:

> Wynne-Wilson developed a way of painting blank slides with brightly coloured inks which became the liquid slides more associated with the London underground scene: he would heat the ink with a small blow-torch and cool it with a hair dryer, causing the ink to bubble and move. The team experimented with different chemicals and colours and each gig revealed new images and effects as they developed their techniques.[69]

An article in the *International Times* in 1966 described the Floyd doing:

> [...] weird things to the feel of the event with their scary feed-back sounds, slide projections playing on their skin – drops of paint run riot on the slides to produce outer space/prehistoric textures on the skin – spotlights flashing on them in time with the drums.[70]

This is a telling description, segueing future and past to describe the alien landscape of light forms creating fleeting textures upon the skin of the musicians as they played. The recurrence of this comparison between ancient and contemporary images – from the Independent Group to the Earth Mysteries movement and into the new medium of the light show – was part of countercultural identity, the desire to reveal new layers of meaning through experience. Just as ley lines could 'reveal' the secrets of an ancient landscape, the abstract patterns of the lightshow could expose relationships between past and present for an audience who were keen to pursue an expanded consciousness. Rycroft has commented upon the role of visual technologies played in this, in '[...] showing new patterns and relationships in nature and the cosmos [they] had a profound effect on how the matter of nature was understood as well as the technologically mediated relationships between humans and nature'.[71] Barry Miles remembered that most of the audience had '[...] never seen a light show before and many stood staring open-mouthed as the amoeba-like bubbles of light pulsated and fused together [...] or expanded and blew apart into dozens of offspring'.[72] The *International Times* columnist Peter Stowell wrote in 1969 that:

> [...] if we agree that between primeval men gazing up at the constellations and audiences watching lightshows, there is only a philosophical difference; then by an examination of origins, developments and ideals we may come to appreciate lightshows as extraordinary vehicles running parallel to man's evolution.[73]

It is the elemental power of light that allows for this mystical interpretation of a technical – and artificial – application of light as entertainment. Stowell also noted that:

> out of the active lightshow [...] the same satisfaction which was once given by the charm of wild inefficient, natural fuel fires and lamps; a satisfaction borne out of wonder and the primeval worship of light. Light is the originator of motion, and by transference, of emotion also."[74]

It is certainly evident that that this thirst for elemental experience expanded into bodily themes in counterculture, from fluids to nudity.

Pete Townshend's Auto-Destruction: anger, feedback and audience

In the dark, arched confines of the Railway Hotel in 1963, The Who were playing to a largely indifferent audience. Feeling frustrated, Townshend sent his guitar launching towards the rafters. This was primarily an effort to create the feedback disturbances with which he had become preoccupied. In an earlier performance this had worked to good effect, but that night Townshend was edgy and angry with frustration. He threw the guitar rather too hard and it broke. The death of the 12-string Rickenbacker certainly gained the audience's attention, to the band's surprise and their manager's delight. Townshend was taken aback but the audience were waiting and watching for more so he made '[...] a big thing of breaking the guitar. I pounced all over the stage with it, threw bits to the crowd, and picked up my spare guitar and carried on as though I really meant to do it.'[75] The sudden violence broke through the barrier between the relatively unknown band and the audience and the response made the band consider how they might manipulate the dynamics of performance to their benefit. Townshend was coerced into repeating his actions and initially refused. Guiliano has noted that the next time, Keith Moon saw off a drum kit at the end of the set, and that 'The antics quickly captured the attention of the press [...] the wily manager not only approached the destruction but encouraged it. From the wings he shouted "Smash your guitar! The *Daily Mail's* here! We can afford it!"'[76]

Since that time, Townshend has placed these early episodes of destruction in performance in the specific context of his art school training. He described the moment he destroyed the guitar:

> On stage I stood on the tips of my toes, arms outstretched swooping like a plane. As I raised the stuttering guitar above my head, I felt I was holding up the bloodied standard of endless centuries of mindless war. Explosions. Trenches. Bodies. The eerie screaming of the wind.[77]

This string of images plays out like the slides that Gustav Metzger had shown to the staff and students of Groundcourse Ealing and Townshend has explicitly linked his actions to Metzger's influence, as Guiliano has recorded:

> Within three weeks, Townshend had latched onto the concept as an art form in the philosophic vein encapsulated by his esteemed Ealing

> College lecturer Gustav Metzke [sic]. 'To me it wasn't violence, it wasn't random destruction,' insists Pete. 'At the time I considered it to be an art. The German movement of autodestructive art. They used to build sculptures that would collapse. They would paint pictures with acid so they autodestructed. They built buildings that would explode.'[78]

As we have touched upon in relation to Groundcourse, nihilistic violence was certainly a visual arts strategy of the era, and this often involved musical instruments too. In 1962, for example Nam June Paik smashed his instrument in *One for Violin Solo* at the Dada music festival in Dusseldorf. Metzger had certainly exposed the students to an alignment between autodestruction and war, producing, through collected slides, a visual mass of nihilistic actions, both military and creative. Townshend has emphasized that 'Metzger proved particularly influential' and that Metzger later '[...] turned up at some [Who] shows when we were smashing stuff up. He really got into it.'[79]

Certainly at Ealing and at most other art schools in the UK, music was a presence that was also transitioning into a mode of experimental practice for staff and students. In 1982, Townshend reminisced that: 'Music was also a big part of life at art school. There would be lessons where you sort of listened to jazz, or listened to classical music, or explored minimalism' and that:

> We had jazz musicians, film writers, playwrights as well as artists [...] to come and lecture. It was a clearing house and music was something that was very much considered to be okay and not something that you only did after hours. It was part of life. You could sit in a classroom with people painting *and* playing. I used to do it.[80]

John Bonehill recalls the high excitement that new music could inspire amongst the students:

> The most important event that happened at Ealing, and which effect was felt around the World, was the day an American student (can't remember his name) convinced the owner of the Cafe across the street from school, to let him put American 'Blues' 45`s on the Juke box. We had no access to this music.[81]

It is little surprise then, that as Mark Wilkerson has observed '[...] Ealing College was a think-tank containing much local artistic talent that went on to musical careers, including future Faces/Rolling Stones guitarist Ronnie Wood and Queen Frontman Freddie Mercury.'[82] This is in addition to Eno and Townshend. Bryan Ferry attended King's College where Ascott trained too. The cross-fertilization between music and art certainly produced some exciting musicians, but it also contributed to the kind of collaborative events outlined here. Michael Craig-Martin, the artist who later taught many of the YBA generation of artists at Goldsmiths, remembers the excitement of British art schools at the time as well as suggesting an alternative reason for artists moving into music:

> I came to Britain in the Sixties, and was immediately teaching in a provincial art school and I couldn't believe how good the school was. They were sophisticated about art. What you often saw were extremely bright students who had failed in the conventional education system. At art college, the things that had been a disadvantage became an advantage. But when people came out of these institutions there was nothing for them. One of the reasons for them going into other fields, including pop music, was that there was nowhere for them to have career in Britain.[83]

The straightforward economic advantages held by the music industry – the large volumes of sales from a single record, the vast popular interest in music and the additional revenue from performances – were seductive. For those trained artists that remained in the visual arts, often a teaching career was the most stable way to augment a practice.

The art school musicians certainly retained an experimental and creative ethos and Townshend has explicitly stated the importance of his art training in his musical career. However Keith Moon had a more straightforward interpretation of the early stage violence performed by Townshend:

> When Pete smashed his guitar, it was because he was pissed off. When I smashed me drums, it was because I was pissed off. We were frustrated. You're working as hard as you can to get that fucking song across, to grab that audience by the balls, to make it an event. When you've done all that, when you've worked your balls off and given the audience everything you can give and they don't give anything back, that's when the fucking instruments go.[84]

He also commented: 'That's one way things got smashed. Another was if a member of the group was too fuckin' stoned to give it their best.'[85] While Moon's interpretation might be less meaningful than Townshend's, this is of little import because the impulse and the action – anger, frustration and violence – were fresh tactics for musicians. This was the beginning of an element of rock music performance that lingers today – like the choreographed violence of wrestlers, the carefully timed destruction of instruments and stage sets at the end of a rock performance is part of the visual code of rock (Figure 7.5). Townshend was later irritated that after watching The Who perform at the Monterey Festival, Jimi Hendrix incorporated smashing his guitar into his performance there too.[86]

One of the interesting elements of the art school rock phenomenon is the ways in which visual arts training influenced modes of performance in

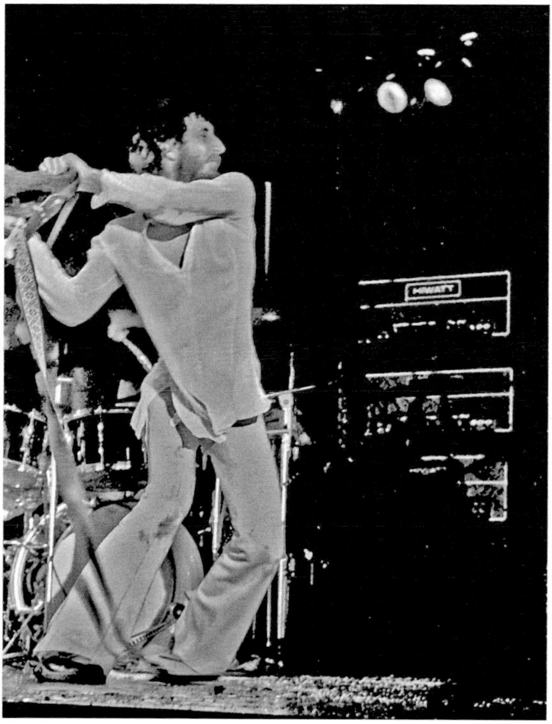

Figure 7.5 Pete Townshend performing with The Who in Hamburg, 1972. Photograph © Heinrich Klaffs

this way. Whether a conscious reference to Metzger or a more general act of subversion, staged nihilistic violence was a new element in the increasingly experimental arena of the rock performance. Moon recalled that these acts of destruction were outbursts of anger and frustration and even this in itself is a marker of a new age; rather than performing to indifference, musicians would try new ways to elicit attention. By 1967, this restless form of protest was a permanent element of the band's performances. In a review of a gig at the Roundhouse in 1967, Nick Jones described the band '[...] winning over the show with an immediate flurry of smoke bombs and sound barrier smashing', before a series of technical faults and holdups provoked Townshend and he 'wheeled upon a fine pair of speakers and ground them with his shattered guitar into the stage'.[87] The next band performing that night was The Move, whose:

> [...] act came smoothly to a stage-shaking climax as TV sets with Hitler and Ian Smith pictures were swiped with iron bars, and a car was chopped up. Two girls were incensed enough to strip to the waist and

> the remaining, shivering crowds surged menacingly towards the stage, the demolished car, and the birds.[88]

The introduction of violence as a performance strategy resulted in a new, subversive energy in rock performance, augmented by the alien screech of feedback and the dark and often claustrophobic performance spaces. Christopher 'Kit' Lambert remembers that when he first saw The Who play at the Railway in 1964: 'The room was black and steaming hot. The audience seemed hypnotized by the wild music, with the feedback Townshend was already producing from his Amp.'[89] Lambert returned the following night with '[...] his partner Chris Stamp', who recalled that he had '[...] never seen anything like it. The Who have a hypnotic effect on the audience. It was like a black mass. Even then Pete was doing all the electronic-feedback stuff [...] the audience were in a trance.'[90]

The comparison to a black mass helpfully links the subversive energy of the period to later, more theatrical references to black magic that were to become such an ingrained element of, for example, heavy metal performance. The countercultural melting pot of folk culture, wiccan magic and spiritualism took on a darker form with distorted sound, anger and nihilistic destruction. Feedback effects were a new and unsettling element of musical performance, often dissonant and jarring with the melodies, creating a more melancholy soundscape. Townshend thought of it in instrumental terms in that:

> You could control it and it could be very musical – certainly that sort of thing where you hit an open A chord and then take your fingers off the strings [...] The A string is still banging away but you're heading the finger-off harmonics in feedback. Then the vibrating starts to stimulate harmonics in the other strings and it's just an extraordinary sound, like an enormous aeroplane. It's a wonderful, optimistic sound and that was something that happened because I was posing – I'd put my arms out, let go of the chord and then find that the resulting noise was better.[91]

Townshend described formless noise, comparing it to the roar of an aeroplane. During their performance at the *International Times* launch, Daevid Allen remembers that Soft Machine '[...] had a motorcycle brought onto stage and would put a microphone against the cylinder head for a good noise'.[92] In both these quotes, it is clear that performance wasn't only about effect or theatricality. Actions such as Allen pressing a mike against a motorbike's engine or Townshend throwing his arms wide to extend a chord were forms of physical experimentation with feedback. Zwerin described Soft Machine's performance, commenting: 'Its cloudy sound moves to unexpected places in weird ways. Everything is filtered through a fuzz box, an electronic gadget that intentionally distorts sound.' He added: 'There is a good deal of collective improvization. Sets are more like suites, each 'tune' running into the next.'[93] In a review of Pink Floyd in the *International Times*, Barry Miles observed that:

> The Floyd were using unconventional techniques: playing guitar with a metal cigarette lighter, rolling ball-bearings down the guitar neck to give an amazing Bo-Diddly feedback sound and feedback in continuous controlled waves, complex repeating patterns that took ages before coming round again'[94]

In each case, playful experiment created unpredictability and excitement in performance, embodying the same ethos of play that was integral to Groundcourse and which, I argue, embodied the reflexivity of the age. This reflexivity also extended into gameplay that drew in the audience – during the Soft Machine set 'Yoko Ono came on stage and created a giant happening by getting everyone to touch each other in the dark, right in the middle of the set'.[95] This incident provoked embarrassed laughter, plunging strangers into intimacy.[96] In a time of increasing physical and sexual liberty, Ono's contribution played upon the reticence the audience felt in transgressing the unspoken boundary of anonymity that binds a group of spectators together; the watchers were drawn into the performance. Indeed, audience inclusion and participation may be the defining feature of countercultural performance, achieved through both musical and conceptual feedback.

Ambience: Eno's cybernetics of sound

There is pleasing symmetry in the fact that one future musician took the experience of Groundcourse and transmuted it into acts of destruction and another moved progressively into ambience. Brian Eno has repeatedly credited his participation in Groundcourse as a vital element of his development. Tom Phillips taught Eno in the early weeks and encountered an original student with a creative and often subversive attitude. In an early life drawing class, Phillips instructed the students to '[...] just make lots of dots as quickly as possible and see if you can say something about the model.[97] Eno made '[...] millions of dots and then tore the whole thing up into the shape of a human being'.[98] It was also Phillips who was to introduce Eno to the music of John Cage, who Phillips claims '[...] made you realise that there wasn't a thing called noise, it was just music you hadn't appreciated'. Apparently, Eno and Brian then developed a game with a few others called 'Sound Tennis'. They '[...] went around Ipswich buying old wrecked pianos and put them all around the room. Then we played a kind of hand tennis and scored according to the quality of noise we made when hitting a stripped piano.'[99]

Eno also engaged with the theoretical content of the course and his interest in cybernetics endured afterwards. In 1995 David Bowie commented:

> Brian is a born cybernetician. He will take the most unlikely juxtapositions and philosophical ideas and throw them together into this kind of

> conceptual stew of his and produce this unfathomable, but fascinating animal. And he will continually stop and re-evaluate the work that's been done and then throw it in an entirely unexpected direction'[100]

There is some truth in Bowie's words – Eno certainly found cybernetics to be an absorbing theoretical model for music and art and he continued to develop this interest long after his studies finished. Pickering has recounted that it was a 'turning point' in Eno's career when, in 1974, Eno's mother-in-law lent him a copy of Stafford Beer's *Brain of the Firm*.[101] Eno then wrote an essay that he sent to Beer, who invited him for a visit. During this visit Beer somewhat oddly offered Eno the torch of cybernetics that had been 'handed to me along a chain from Ross Ashby, it was handed to him from [...] Warren McCulloch'.[102]

Several writers have credited Stafford Beer's influence upon Eno as a crucial one. Sytze Steenstra commented that his '[...] approach to musical composition is fundamentally influenced by cybernetic management theory' and that he '[...] applies principles of decision and control that have been developed by cybernetic thinkers'.[103] Pickering offers a compelling account of the place of cybernetics in Eno's music, summing up his position on Beer as follows: 'Beer's ontology of exceedingly complex systems conjures up a lively world, continually capable of generating novel performances. Eno [...] picked up the other end of the stick and focused on building musical worlds that would themselves exhibit unpredictable, emergent becomings.'[104] This idea echoes Eno's own words in the liner of his 1975 album *Discreet Music*, when he wrote:

> Since I have always preferred making plans to executing them, I have gravitated towards situations and systems that, once set in operation, could create music with little or no intervention on my part. That is to say, I tend towards the roles of planner and programmer, and then become an audience to the results.[105]

It is worth noting, first, that these ideas were not simply extracted from Stafford Beer's writing but were more generally applicable to other cybernetic theories too and, second, this mode of systems-oriented working was common in music, theatre and the visual arts. Ascott had employed a similar principle in his hiring policy of 'requisite variety', choosing staff from varied disciplines and with very different approaches and then allowing the course to 'perform' its function in unpredictable ways.[106] Focusing on the system not the outcome in this manner was certainly connected to the cybernetic ethos of allowing the system to 'perform'. In this sense, playful experiment is transdisciplinary but of cybernetic origin. It is, however, worth emphasizing the role that Groundcourse and other pedagogical models held in sharing and popularizing cybernetics as an art strategy. Groundcourse staff and art educators across the country also introduced their students to music, including the work of John Cage. The experimental musicians Morton Feldman and Christian Wolf were amongst a number of visiting staff and speakers at Ipswich.[107] Christina Dunbar-Hester has commented that:

> [...] the resonance between ideas undergirding experimental and electronic music composition and construction in the 1950s–1970s and cybernetic theories may be the result of prevalent, complex, and interrelated ideas about human-machine interaction and relationships, ideas about communication and control, and changing aesthetics.[108]

This statement could equally describe visual art in the same period. The many concurrent threads of discussion about cybernetics in relation to art music and literature together build a picture that extends beyond any single discipline. Indeed, the many multimedia events and performances detailed in this chapter were transdisciplinary rather than interdisciplinary, exploring feedback both with instruments and equipment and through interactive and participatory performance.

However, Eno was unusually direct in his response to cybernetics, exhibiting a detailed understanding of cybernetic concepts and an experimental application of these concepts within his practice. In the lyrics of *Seven Deadly Finns* of 1973, he wrote:

> Although variety is the spice of life
> A steady rhythm is the source
> Simplicity is the crucial thing
> Systematically of course
> (work it all out like Norbert Wiener)[109]

Two years previously, the artist Hans Haacke had aborted a project intended for the Guggenheim named *Norbert: All Systems Go*. He had tried to train a mynah bird in order to achieve the following proposal:

> bird speaks. 'All systems go' it squawks. And again, 'All systems go.' A pause. 'All systems go. All systems go.' Repetition to inanition. 'All systems go.'[110]

In each case Norbert Wiener occupied a cult position, a figurehead for systems behaviour. Haacke's project, had it come off, would have been unpredictable given the living element; the mynah bird was a black box, which could only be understood through performance and observation. Similarly, Eno wanted to design systems that would then perform unpredictably within a set of considered perimeters. Far from the theatrical use of feedback and audience response that fuelled the early success of The Who, Eno moved progressively toward a more democratic method that as Steenstra described, '[...] would be anti-hierarchical [...] using the studio as an instrument to generate new musical hybrids. As a result, Eno used the recording studio in a manner that was the reverse of a dominant trend in 70s rock.'[111]

It is worth noting that the concept of 'studio as instrument' had evolved painstakingly over the previous decades in step with the development of the technologies which make up a recording studio. As Ralph Parkman has noted: 'Throughout the 1950s, musique concrete and electronic music were active areas of research but the actual physical processes were outstandingly tedious. A two or three minute composition could take months of patient work.' He added: 'Of all the arts, it seems that music stands to be carried furthest by the research with computers [...].'[112]

In 1976, Eno's interest in the creative possibilities afforded by the music studio coalesced into a new musical concept. He wrote:

> My contention is that a primary focus of experimental music has been toward its own organization, and toward its own capacity to produce and control variety, and to assimilate "natural variety" – the "interference value" of the environment.[113]

Two years later he was to 'release the first record which described itself as ambient music'.[114] Eno had observed changed listening habits amongst his friends, who were seeking '[...] stillness, homogeneity, lack of surprises, and most of all, lack of variety'.[115] Steenstra described this as an '[...] an atmospheric or environmental music without defined focus'.[116] Eno also wanted to demonstrate what to him had been evident for some time – that studios, with all their knobs and dials, were not the domain of engineers, but of musicians and that it was 'actually one of the distinguishing characteristics of new music [...]'.[117] Ambience, then, is a decentred musical practice that resists predictable form, creating atmosphere and mood. Eno wrote that 'A musical score is a statement about organization; it is a set of devices for organizing behaviour toward producing sounds', a definitively cybernetic description of musical composition as system.[118] This application of cybernetic thinking to music production has a detached elegance that is quite at odds with the wild and subversive experiments in musical feedback that had shaken up the music scene a decade earlier. Just as early cyberneticians playfully experimented with messy and creative machines before moving onto more formal experiments as technologies evolved, so did musicians.

Synthesis

Groundcourse students dispersed into the British art scene with an unusually networked view of art practice as situated in larger systems of production and exchange. This was augmented by an understanding that the new age of communication technology, while in its infancy, would transform the world in the decade ahead. However, there is also one inheritance from Groundcourse that is less easily quantified. For Ascott and for the students that he taught, technology was not just a tool but a site for creative practice in itself. The curious magic of the signals, readings and sounds as transmission of information created new opportunities for artists, whether it was utilizing technologies directly or experimenting with feedback as a model of participatory practice. While from the contemporary perspective the curriculum of Groundcourse appears radical, it was situated in perhaps the most important contemporary debates of its time: technology, communication and human behaviour. As I have argued, these new areas of research and practice were rooted in World War II, but in the post-war years, they became resolutely future-facing.

It is also worth noting that the collaborative work model necessitated by war – between engineers, mathematicians, biologists and psychiatrists – generated a fruitful period of interdisciplinary exchange in the post-war period. In this sense, the period of collaboration that extended into the 1970s originated with cybernetics. Its simultaneous relevance in the military (logistics, weaponry, strategy), society (organizational management, computing, education) and the arts (interdisciplinary projects, participatory tactics, multimedia) together constituted the spirit of the age.

The multi-layered convergence of art forms and technologies that formed British countercultural happenings were, in essence, cybernetic. That is to say, the unifying concept of feedback draws the strands of sound, light, art, performance and participation together, into a cybernetic system. Multimedia is an insufficient term for this phenomenon for several reasons. Firstly, the term focuses on the medium instead of the message – that is, it brings to mind the apparatus of film, sound and light rather than the collective outcomes of these arrangements of media. Secondly, it does not encapsulate the other layers of activity that took place, including costume, performance art, dance, sculpture and painting. Finally, it does not bring to life the way that these events existed *in time*: that is, as complex systems of interaction that created experiences that only existed for the duration of an event. The term 'happenings' effectively frames these events in time, but it does little to describe the vital systems element of these events. Willoughby Sharp's little-used term 'total integration' captures this better.[119]

Ascott's Groundcourse drew together the strands of technology, behaviour and communication and through live pedagogical experiment involving staff and students, he created a total work of art that stands as an exemplar of the many points of overlap between cybernetics and art practice in the 1960s. It brings to life the cybernetic precept that communication is something that happens *between* or *through*; it is not located in a fixed object (such as a painting or sculpture) but instead exists through interaction. This concept unifies many of the disparate and experimental practices that evolved in the visual arts in the two immediate post-war decades. In the decades that have intervened, Ascott has retained the same cybernetic principles in his practice. For Ascott, cybernetics never ended as such; it simply dispersed into other applications and theoretical models. Dispersal and synthesis then – of ideas, practices, media and people – were defining conditions of this very rich period for British visual culture.

Notes

1 Simon Frith & Howard Home (1987) *Art into Pop*. London: Routledge, 1.
2 Frith & Home (1987).
3 Kate Sloan *Interview with Keith Albarn* (by telephone, 2018).
4 John Bonehill (2011) Email Exchange with the author.
5 Gorgio Gomelsky. Email to Aymetric Laroy. Quoted in a blog article on the Kevin Ayers website: Susan Lomas (2015) 'Do you know the Way to St Tropez?' Available online at http://www.kevinayers.org/tours/st-tropez, accessed 12 November 2017.
6 John Bonehill (2011) Email Exchange with the author.
7 *Ibid.*
8 Martin Roach & David Nolan (2015) *Damon Albarn – Blur, Gorillaz and Other Fables*. London: John Blake Publishing.
9 Susan Lomas attributes this quote to *L'Express Magazine*, 3–6 July 1967.
10 Graham Bennett (2001) *Soft Machine: Out-bloody-rageous*. London: SAF Publishing, 108.
11 Keith Albarn email to Aymetric Laroy, quoted in Lomas (2015).
12 Kate Sloan *Interview with Keith Albarn* (by telephone, 2018).
13 Bonehill quoted in Lomas (2015).
14 Mike Zwerin (1968) 'The Soft Machine' *Downbeat*, July 11. Available online at http://www.disco-robertwyatt.com/images/Robert/interviews/DownbeatJuly1968/index.htm, accessed 11 November 2017.
15 *Ibid.*
16 Elda Danese 'Soft Machine' in: James Everett Katz (ed.) (2003) *Machines that Become Us: The Social Context of Personal Communication Technology*. New Jersey: Transaction Publishers, 267–76, 267.
17 Zwerin gives an account of this, describing how when they toured the US, Pink Floyd had a salary of $100 a week, but they paid Boyle out of it too, causing friction with their management who did not appreciate the value of the light show. See Zwerin (1968).
18 John Bonehill (2011) Email Exchange with the author.
19 Mark Boyle. In: J.L. Locher (1978) *Mark Boyle's Journey to the Centre of the Earth*. Stuttgart & London: Edition Hansjörg Myer, 76.
20 Mike Ratledge quoted in Zwerin (1968).
21 John Bonehill in Lomas (2015).
22 *Ibid*. 79.
23 *Ibid*. 79.
24 Herbert Read (1947) 'The Fate of Modern Painting'. *Horizon*, November, 242–54, 253.
25 John Bonehill quoted in Lomas (2015).
26 Daevid in: Barry Miles (2011) *Pink Floyd: The Early Years*. London: Omnibus Press, 79.
27 *Ibid*. 79.
28 Simon Rycroft (2011) *Swinging City: A Cultural Geography of London 1950–1974*. London: Routledge, 153.
29 *Ibid*., 8.
30 *Ibid*., 8.
31 David Mellor (1993) *The Sixties Art Scene in London*. London: Phaidon Press & Barbican Gallery, 185.
32 Mellor (1993) 185.
33 Jasia Reichardt (1973) 'Art at Large', *New Scientist*. 8 February, 327.
34 Zwerin (1968).

35 Mark Boyle quoted in: M G McNay (1967) 'Son et Lumiere at the Sandon Theatre, Liverpool'. *The Guardian*, Jan 11, 9.
36 Mark Boyle & Joan Hills 'Son et Lumiere for Bodily Fluids and Functions' in J.L. Locher (1978) *Mark Boyle's Journey to the Centre of the Earth*. Edition Hansjorg Myer 125.
37 Daevid Allen quoted in: Miles (2011) 79.
38 M G McNay (1967) 'Son et Lumiere at the Sandon Theatre, Liverpool' *The Guardian*, Jan 11, 9.
39 Mark Boyle in J.L. Locher (1978) 72.
40 *Ibid.*
41 Joel Brown 'I was there' interviewed by Holly Lucas (youth oral historian) for The Roundhouse, 15 October 2016. Recording. Available online at https://50.roundhouse.org.uk/content-items/opening-night-lighting-designer-joel-brown, accessed on 11 December 2017.
42 Julian Palacios (2010) *Syd Barratt & Pink Floyd: Dark Globe* London: Plexus Publishing Ltd. (digital version unpaginated).
43 Brown & Lucas (2016).
44 *Ibid.*
45 *Ibid.*
46 *Ibid.*
47 Nick Jones (1967) 'The Who, The Move, Pink Floyd: The Roundhouse, Chalk Farm, London'. *Melody Maker*, 7 January.
48 *Ibid.*
49 *Ibid.*
50 Daevid Allen quoted by Miles (2011) 79.
51 See 'Field' chapter for further discussion of the collision of ancient myth and science fiction in British counterculture. Michell (1969).
52 Zwerin (1968).
53 *Ibid.*
54 Palacios (2010) unpaginated.
55 Mellor (1993) 185.
56 Rycroft (2011) 150.
57 Miles (2011) 79–80.
58 *Ibid.*, 80.
59 Andrew Pickering (2011) *The Cybernetic Brain: Sketches of Another Future* Chicago: University of Chicago Press, 206.
60 *Ibid.*, 207.
61 R.D. Laing 'Metanoia: Some Experiences at Kingsley Hall, London'. In: H. M. Ruitenbeek (ed.) (1972) *Going Crazy*. New York: Bantam, 11–21, 12.
62 *Ibid.*, 12.
63 *Ibid.*, 16.
64 Mellor (1993) 185.
65 Miles (2011) 107.
66 Miles (2011) 108.
67 Rycroft (2011) 154.
68 Mark Boyle in *Tune In, Turn On, Light Up Tate Etc.*, Summer 2005, Issue 4. 70–2.
69 Miles (2011) 68.
70 I.T. Article quoted in Rycroft (2011) 153.
71 Rycroft (2011) 153.
72 Miles (2011) 73.
73 Peter Stowell (1969) 'Head-Lights' *International Times*, Vol. 54, 13.
74 *Ibid.*, 13.

75 Townshend quoted by Guiliano (2002) 53.
76 Guiliano (2002) 53.
77 Pete Townshend in Casey Harison (2017) *Feedback: The Who and their Generation*. London: Rowman and Littlefield, 103.
78 Guiliano (2002) 53.
79 Wilkerson (2009) *Who Are You: The Life of Pete Townshend*. London: Omnibus Press, 12.
80 *Ibid.*, 12.
81 John Bonehill (2011) Email exchange with the author.
82 Wilkerson (2009) 11.
83 Michael Craig-Martin quoted in Wilkerson (2009) 12.
84 Guiliano (2002) 54.
85 Guiliano (2002) 54.
86 *Ibid.*, 78.
87 Nick Jones (1967) 'The Who, The Move, Pink Floyd: The Roundhouse, Chalk Farm, London' *Melody Maker*, 7 January.
88 *Ibid.*
89 Guiliano (2002) 51.
90 Guiliano (2002) 51.
91 Wilkerson (2009) 19.
92 Miles (2011) 76.
93 Zwerin (1968).
94 Miles (1967) *International Times*.
95 Miles (2011).
96 Palacios (2010) unpaginated.
97 Tom Phillips in: Geeta Dayal (2009) *Brian Eno's Another Green World*. London: A & C Black, 16.
98 *Ibid.*, 16.
99 *Ibid.*, 17.
100 Christina Dunbar-Hester (2010) 'Listening to Cybernetics: Music, Machines, and Nervous Systems'. *Science, Technology, and Human Values*, Vol. 35. No. 1, 113–37.
101 Pickering (2011) 11.
102 Brian Eno *Ibid.*, 303.
103 Sytze Steenstra (2010) *Song and Circumstance: The Work of David Byrne from Talking Heads to the Present*. London: Continuum, 8.
104 Pickering (2011) 304.
105 Brian Eno, liner notes to 1975 album *Discreet Music*.
106 Kate Sloan *Interview with Roy Ascott* (September 2011).
107 Roy Ascott in Bracewell (2011) 202.
108 Dunbar-Hester (2010) 115.
109 Brian Eno (1973) *Seven Deadly Finns*.
110 For an account of this project see Luke Skrebowski (2006) 'Systems Aesthetics' *Tate Papers*, no. 5, Spring. Available online at http://www.tate.org.uk/research/publications/tate-papers/05/all-systems-go-recovering-jack-burnhams-systems-aesthetics, accessed January 2017.
111 Steenstra (2010) 8.
112 Ralph Parkman (1972) *The Cybernetic Society*. New York: Pergamon Press Inc., 303.
113 Brian Eno (1976) 'Generating and Organizing Variety in the Arts'. Reproduced in: Christoph Cox & Daniel Warner (eds) (2017) *Audio Culture: Readings in Modern Music*. London: Bloomsbury 315–25, 316.
114 Brian Eno 'Ambient Music' *Ibid.*, 109–12, 109.

115 *Ibid*., 110.
116 Steenstra (2010) 8.
117 Eno (1976) 110.
118 Eno (1976) 315.
119 Willoughby Sharp (1967) 344.

Index

For Product Safety Concerns and Information please contact our EU representative GPSR@taylorandfrancis.com Taylor & Francis Verlag GmbH, Kaufingerstraße 24, 80331 München, Germany